FOREWORD

This volume contains the texts of the lectures and seminars given at the C.I.M.E. Session on Schrödinger Operators, held at the Centro di Cultura Scientifica "A.Volta", Villa Olmo, Como, Italy, from August 27 to September 4, 1984.

The objective of the Session was to provide a broad and up-to-date survey of the meeting-ground common to functional analysis, partial differential equations and quantum mechanics, which goes under the name of Schrödinger operator theory. The three main courses delivered include a thorough coverage of two specific research areas, namely eigenfunctions estimates for the n-body problem by Professor S.Agmon and asymptotic completeness for three-body scattering by Professor V.Enss, and a general overview of basic and recent results by Professor B.Simon. The seminars cover three rapidly expanding research subjects: transition to chaos (J.Bellissard) and probabilistic techniques (G. Jona-Lasinio) in quantum mechanics, and the classical limit by the technique of Fourier integral operators (K.Yajima).

The editor is deeply grateful to his colleagues Professor V.Grecchi and Dr. E.Caliceti and to the staff of the Centro di Cultura Scientifica A.Volta for their assistance in the organization of the Session.

Bologna, March 1985

Sandro Graffi

TABLE OF CONTENTS

C.I.M.E. Session on Schrödinger Operators

List of Participants

S. Agmon, Inst. of Math. and Comp. Sci., The Hebrew Univ., Givat Ram,
91904 Jerusalem, Israel

J. Asch, Yorchstr. 72, D-1000 Berlin 61

J. Bellissard, C.N.R.S., Centre de Physique Théorique, Luminy-Case 907,
13288 Marseille Cedex 9, France

P. Biler, Institute of Mathematics, Univ. of Wrocław, p. Grunwaldzki 2/4,
50-384 Wrocław, Poland

A. Bove, Dipartimento di Matematica, Università di Trento, 38050 Povo, Trento

P. Briet, C.N.R.S., Centre de Physique Théorique, Luminy-Case 907,
13288 Marseille Cedex 9, France

E. Caliceti, Istituto Matematico Università, Via Campi 213/B, 41100 Modena

J. van Casteren, Rendierstraat 15, 2610 Wilrijk, Belgium

A.M. Charbonnel, Inst. de Math. et d'Inf., Univ. de Nantes, 2 rue de la Houssinière,
44072 Nantes Cedex, France

J.M. Combes, Univ. de Toulon, 83130 La Garde, France

M. Combescure, LPTHE Batîment 211, Univ. de Paris-Sud, 91406 Orsay, France

H. Cycon, TU Berlin, Fachbereich Mathematik, Strasse d. 17 Juni 135, 1 Berlin 72

E.M. Czkwianianc, Instytut Matematyki Uniwersytety Lodzkiego, ul. S. Banacha 22,
90-239 Łódz, Poland

A. Devinatz, Dept. of math., Northwestern Univ., Evanston, IL 60201, USA

J.F.R. Donig, Technische Hochschule Darmstadt, Fachbereich Mathematik,
Arbeitsgruppe 12, Schlossgartenstr. 7, D-6100 Darmstadt

G. Dore, Dipartimento di Matematica, Piazza di Porta S.Donato 5, 40127 Bologna

P. Duclos, C.N.R.S., Centre de Physique Théorique, Luminy-Case 907,
13288 Marseille Cedex 9, France

B. Ducomet, CEREMADE, Place du Maréchal de Lattre de Tassigny,
75775 Paris Cedex 16, France

D. Eidus, School of Math. Sciences, Tel Aviv Univ., Ramat Aviv,
69978 Tel-Aviv, Israel

V. Enss, Freie Univ. Berlin, Fachbereich Math., WE 1 Arnimallee 3, 1000 Berlin 33

C. Erdmann, Technical Univ. Berlin, Dept. of Math., Strasse des 17 Juni 136,
1000 Berlin 12, FRG

J.E.G. Farina, Dept. of Math., Univ. of Nottingham, University Park,
Nottingham NG7 2RD, England

J. Fleckinger, 41 rue Boyssonne, 41400 Toulouse, France

G. Fonte, Istituto Dipartimentale di Fisica, Corso Italia 57, 95125 Catania

F. Franchi, Dipartimento di Matematica, Piazza di Porta S. Donato 5, 40137 Bologna

M. Frasca, Via G. Vagliasindi 38, 95126 Catania

N. Garofalo, School of Mathematics, Univ. of Minnesota, 127 Vincent Hall, Minneapolis, Minn. 55455, USA

J.M. Ghez, C.N.R.S., Centre de Physique Théorique, Luminy - Case 907, 13288 Marseille Cedex 9, France

S. Graffi, Dipartimento di Matematica, Piazza di Porta S. Donato 5, 40127 Bologna

V. Grecchi, Dipartimento di Matematica, Università di Modena, Via Campi 213/B, 41100 Modena

G. Hagedorn, Dept. of Math., Virginia Polytechnic Institute and State Univ., Blacksburg, Virginia 24061, USA

E. Harrell, School of Mathematics, Georgia Institute of Technology, Atlanta, GA 30332-0160, USA

Y. Herczynski, Institute of Mathematics, Warsaw University PkIN, Warsaw 00901,Poland

M. Hoffmann-Ostenhof, Institut fur Theoret. Physik d. Universitat Wien, Boltzmanngasse 5, 1090 Wien, Austria

T. Hoffmann-Ostenhof, Inst. fur Theoret. Chemie u. Strahlenchemie d. Universitat Wien, Wahringerstr. 17, 1090 Wien, Austria

W. Hunziker, Institut fur Theor. Physik, ETH-Hoenggerberg, CH-8093 Zurich

B. Johansson, Math. Dept., Chalmers Univ. of Technology and Goteborgs University, S-41296 Goteborg, Sweden

G. Jona-Lasinio, Univ. di Roma "La Sapienza", Dipartimento di Matematica, P.le Aldo Moro, 5, 00185 Roma

G. Karner, Fakultat fur Physik, Univ. Bielefeld, D-4800 Bielefeld 1, FRG

W. Kirsch, Inst. fur Mathematik, Ruhr Universitat, D-4630 Bochum, W. Germany

M. Krishna, Indian Statistical Institute, 7 S.J.S. Sansanwal Marg, New Delhi - 110016, India

B. Lascar, 36 rue du Pré St. Gervais, 93500 Pantin, France

R. Lascar, Université Paris VII, Dept. de Math., 2 Place Jussieu, Paris 5e, France

J.E. Lewis, Dipartimento di Matematica, Università di Bologna, Piazza di Porta S. Donato 5, 40127 Bologna

C. Macedo, Fakultat fur Physik, Univ. Bielefeld, D-4800 Bielefeld 1, West Germany

M. Maioli, Dipartimento di Matematica, Università di Modena, Via Campi 213/B, 41100 Modena

G. Mantica, Dipartimento di Fisica, Sez. Fisica Teorica Appl., Università di Milano, Via Celoria 16, 20133 Milano

P. Maslanka, Inst. of Math., Univ. of Łódz, ul. St.Banacha 22, 90-238 Łódz,Poland

G. Modica, Istituto di Matematica Applicata, Via S. Marta 3, 50139 Firenze

S. Nakamura, Omiya-shi, Miya-machi 3-135, Saitama, Japan T330

F. Nardini, Via Marconi 9, 40122 Bologna

C. Nessmann, Theoretische Physik, Fakultat fur Physik, Univ. Bielefeld, D-4800 Bielefeld 1, FRG

A. Outassourt, 8 rue de l'Hotel de Ville, 44000 Nantes, France

T. Paul, Centre de Physique Théorique, C.N.R.S. Luminy, Case 907, 13288 Marseille Cedex 9, France

M. Perusch, Institut fur Theoretische Physik, Univ. Graz, Universitatsplatz 5, A-8010 Graz, Austria

P. Piccoli, International School for Advanced Studies (SISSA), Strada Costiera 11, 34014 Miramare Grignano, Trieste

Y. Pinchover, Institute of Mathematics, Hebrew University, Givat Ram, Jerusalem 91904, Israel

L. Pittner, Institut fur theoretische Physik, Univ. Graz, Universitatsplatz 5, A-8010 Graz, Austria

W. Plass, Freie Univ. Berlin, Fachbereich Mathematik, Arminallee 2-6, 1 Berlin 33

A. Raphaelian, Sekr. MA 7-1, TU Berlin, Str. des 17 Juni 136, D-1000 Berlin 12, West Germany

D. Robert, 3 Allée M. Croz, 44300 Nantes, France

M. Serra, Via Mattia Farina 2, 84100 Salerno

I.M. Sigal, Dept. of Theoretical Mathematics, The Weizmann Inst. of Sci., Rehovot, Israel 76100

H. Silverstone, Dept. of Chemistry, The Johns Hopkins Univ., Baltimore, Maryland 21218, USA

B. Simon, CALTECH, Mathematics, Pasadena, CA 91125, USA

G. Stolz, Burgstr. 16, 6457 Maintal 4, Germany

R. Svirsky, Dept. of Math., The Johns Hopkins Univ., Baltimore, MD 21218, USA

B. Thaller, Institut fur Mathematik I, Freie Univ. Berlin, Arnimallee 3, D-1000 Berlin, 33

A. Venni, Dipartimento di Matematica, Piazza di Porta S. Donato 5, 40127 Bologna

A. Voros, Service de Physique Théorique, CEN Saclay, F-91191 Gif-sur-Yvette, France

J. Walter, Inst. fur Mathematik, Templergraben 55, D-5100 Aachen

D. White, Dept. of Mathematics, Univ. of British Columbia, Vancouver, B.C. V6T 1W5, Canada

U. Wuller, Sonnenallee 54, 1000 Berlin 44

K. Yajima, Dept. of Pure and Applied Sciences, Univ. of Tokyo, 3-8-1 Komaba, Meguro-ku, Tokyo 153 Japan

Bounds on exponential decay of eigenfunctions of Schrödinger operators

by

Shmuel Agmon

Introduction

Consider a Schrödinger differential operator $P = -\Delta + V$ on $\mathbb{R}^n$ where V is a real function in $L^1_{loc}(\mathbb{R}^n)$. Under general conditions on V the operator P admits a unique self-adjoint realization in $L^2(\mathbb{R}^n)$ which we denote by H. Set

$$\Sigma = \inf \sigma_{ess}(H)$$

where $\sigma_{ess}(H)$ is the essential spectrum of H (assumed to be bounded from below).

It is known that for a general class of potentials V any eigenfunction of H with eigenvalue situated in the discrete spectrum decays exponentially. In these lectures we shall study in some detail the pattern of decay of such eigenfunctions with eigenvalues situated below the bottom of the essential spectrum. We shall discuss two possible methods to derive bounds on eigenfunctions. The methods use two different positivity properties of the operator $P-\lambda$ for λ below the bottom of the essential spectrum.

The first method relies on the fact that for $\lambda < \Sigma$ the equation $(P-\lambda)u = 0$ admits positive supersolutions in some neighborhood of infinity in $\mathbb{R}^n$ and that these yield upper bounds on eigenfunctions.

The second method yields upper bounds on eigenfunctions by exploiting the fact that for $\lambda < \Sigma$ the quadratic form $((P-\lambda)\varphi,\varphi)$ is positive on $C_o^\infty(\mathbb{R}^n \setminus K)$ for some compact set $K \subset \mathbb{R}^n$. More precisely, one considers quadratic form lower bounds of the form:

$$((P-\lambda)\varphi,\varphi) \geq \int_{\mathbb{R}^n} c(x)\,|\varphi|^2 dx \tag{1}$$

for all $\varphi \in C_o^\infty(\mathbb{R}^n \setminus K)$ where $c(x)$ is some positive continuous function on $\mathbb{R}^n$. Under some general conditions it is shown that if ψ is an eigenfunction of H with eigenvalue λ then (1) implies the upper bound:

$$\psi(x) = O(e^{-(1-\varepsilon)\varrho(x)}) \quad \text{as} \quad |x| \to \infty$$

for any $\varepsilon > 0$ where $\varrho(x)$ is the geodesic distance from x to the origin in $\mathbb{R}^n$ in the Riemannian metric:

$$ds^2 = c(x)|dx|^2 .$$

The plan of these lectures is as follows. After some preliminaries we discuss in section 1 the self-adjoint realization of P in $L^2(\mathbb{R}^n)$, establishing among other things Persson's formula for the bottom of the essential spectrum. In section 2 we discuss positive solutions and supersolutions of the equation $(P-\lambda)u = 0$ in a neighborhood of infinity in $\mathbb{R}^n$ and prove that these yield majorants for L^2 solutions. In section 3 we apply the results of section 2 to derive precise upper and lower bounds on eigenfunctions of two body Schrödinger operators. In section 4 we derive weighted L^2 estimates for solutions of $(P-\lambda)u = 0$ in a neighborhood of infinity assuming the validity of quadratic form lower bounds such as (1). Finally, after some preparations in section 5, we apply the results in section 6 to the study of exponential decay of eigenfunctions of N-body type Schrödinger operators.

0. Preliminaries

We introduce some notation and definitions.

We denote by Ω an open set in $\mathbb{R}^n$. We denote by $B_R(x)$ the open ball in $\mathbb{R}^n$ centered at $x = (x_1,\dots,x_n)$ and with radius R. We set $\Omega_R = \mathbb{R}^n \setminus \overline{B_R(0)}$. We let ∂_i denote $\partial/\partial x_i$.

$H^1(\Omega)$ denotes the Sobolev space of functions $u \in L^2(\Omega)$ with distributional derivatives $\partial_i u \in L^2(\Omega)$ for $i = 1,\dots,n$. $H^1_{loc}(\Omega)$ is defined similarly with respect to $L^2_{loc}(\Omega)$.

We shall consider in these lectures solutions of the Schrödinger equation:

$$-\Delta u + (V(x) - \lambda)u = f(x) \quad \text{in } \Omega \tag{0.1}$$

where Δ is the Laplacian on $\mathbb{R}^n$, V is a real function in $L^1_{loc}(\Omega)$, λ a constant and f is a given function in $L^1_{loc}(\Omega)$. By a solution u of (0.1) we shall always mean a weak solution of the following form: u is a function in $H^1_{loc}(\Omega)$ with $Vu \in L^1_{loc}(\Omega)$ such that

$$\int_\Omega (\nabla u\cdot\nabla\varphi + (V-\lambda)u\varphi)\,dx = \int_\Omega f\varphi\,dx \tag{0.2}$$

for every $\varphi \in C_0^\infty(\Omega)$.

We shall also consider supersolutions and subsolutions of (0.1) when λ and f are real. Thus a supersolution of (0.1) is a real function u in $H^1_{loc}(\Omega)$ with $Vu \in L^1_{loc}(\Omega)$ such that

$$\int_\Omega (\nabla u \cdot \nabla \varphi + (V-\lambda) u\varphi)\, dx \geq \int_\Omega f\varphi\, dx \tag{0.3}$$

for every non-negative $\varphi \in C_o^\infty(\Omega)$.

On certain occasions we shall use the following results on solutions of the homogeneous equation:

$$-\Delta u + (V-\lambda) u = 0 \quad \text{in} \quad \Omega . \tag{0.4}$$

<u>0.1 Theorem</u>: Suppose that $V \in L^p_{loc}(\Omega)$ with $p > n/2$, then

(i) All solutions of (0.4) are Hölder continuous.
(ii) Non-negative solutions of (0.4) verify Harnack's inequality.
That is, given a connected compact set $K \subset \Omega$, there exists a constant C such that if u is a non-negative solution of (0.4), then

$$\max_K u \leq C \min_K u \tag{0.5}$$

For a proof of Theorem 0.1 in a more general form see [7].

The following result is well known (e.g. [6; p. 302]).

<u>0.2 Lemma</u>: Suppose that $g \in L^p(\mathbb{R}^n) + L^\infty(\mathbb{R}^n)$ with $p > n/2$. Then for every $\varepsilon > 0$ there exists a constant C_ε depending only on ε , p and n such that

$$\int_{\mathbb{R}^n} |g||\varphi|^2 dx \leq \varepsilon \int_{\mathbb{R}^n} \sum_{i=1}^n |\partial_i \varphi|^2 dx + C_\varepsilon \int_{\mathbb{R}^n} |\varphi|^2 dx \tag{0.6}$$

for all $\varphi \in C_o^\infty(\mathbb{R}^n)$.

<u>1. Self-adjoint realizations of Schrödinger operators</u>

In this section we consider the problem of the self-adjoint realization of the operator:

$$P = -\Delta + V \tag{1.1}$$

in $L^2(\mathbb{R}^n)$ where V is a real function in $L^1_{loc}(\mathbb{R}^n)$.
Set: $V_- = \max(0,-V)$. We impose the following

Condition on V_- : There exist constants $0<\theta<1$ and $C>0$ such that

$$\int_{\mathbb{R}^n} V_-|\varphi|^2 dx \leq \theta \int_{\mathbb{R}^n} |\nabla\varphi|^2 dx + C\int_{\mathbb{R}^n} |\varphi|^2 dx \tag{1.2}$$

for all $\varphi \in C_0^\infty(\mathbb{R}^n)$ where $|\nabla\varphi|^2 = \sum_{i=1}^n |\partial_i\varphi|^2$.

Note that in view of Lemma 0.2 a sufficient condition for (1.2) to hold is that $V_- \in L^p(\mathbb{R}^n) + L^\infty(\mathbb{R}^n)$ for some $p > n/2$. For a weaker condition which ensures (1.2), see [1 ; Lemma 0.3].

We define

$$(Pu,v) = \int_{\mathbb{R}^n} (\nabla u\cdot\nabla\bar{v} + Vu\bar{v})\, dx$$

for any two functions u and v in $H^1_{loc}(\mathbb{R}^n)$ for which the last integral exists.

It is clear that

$$(P\varphi,\varphi) = \int_{\mathbb{R}^n} (|\nabla\varphi|^2 + V|\varphi|^2)\, dx \leq \int_{\mathbb{R}^n} (|\nabla\varphi|^2 + |V|\,|\varphi|^2)\, dx \tag{1.3}$$

for all $\varphi \in C_0^\infty(\mathbb{R}^n)$. Moreover, using (1.2) it follows readily that there exists a constant $C_0 > 0$ such that

$$(P\varphi,\varphi) + C_0\|\varphi\|^2 \geq \delta \int_{\mathbb{R}^n} (|\nabla\varphi|^2 + |V|\,|\varphi|^2)\, dx \tag{1.4}$$

for all $\varphi \in C_0^\infty(\mathbb{R}^n)$ where $\delta = (1-\theta)/2$. Here and in the following $\|\cdot\|$ denotes the norm in $L^2(\mathbb{R}^n)$.

Under the above conditions the main result on self-adjointness is given in the following

1.1 Theorem: Define an operator $H : L^2(\mathbb{R}^n) \to L^2(\mathbb{R}^n)$, with domain $D(H) \subset L^2(\mathbb{R}^n)$, as follows:

$$D(H) = \{u:\ u \in H^1(\mathbb{R}^n),\ |V|^{1/2}u \in L^2(\mathbb{R}^n),\ Pu \in L^2(\mathbb{R}^n)\}$$

$$Hu = Pu \quad \text{for} \quad u \in D(H) \tag{1.5}$$

where P acts on u in the distribution sense.

Then H is a self-adjoint operator bounded from below. Also, if $H_\lambda = (H-\lambda) \geq 0$ for $\lambda \in \mathbb{R}$, then

$$D(H_\lambda^{1/2}) = \{u:\ u \in H^1(\mathbb{R}^n),\ |V|^{1/2}u \in L^2(\mathbb{R}^n)\} \tag{1.6}$$

<u>Proof</u>: By adding to P a constant we may assume with no loss of generality (using (1.4)) that

$$(P\varphi,\varphi) \geqslant \|\varphi\|^2 + \delta\int_{\mathbb{R}^n}(|\nabla\varphi|^2 + |V|\,|\varphi|^2)\,dx \tag{1.7}$$

for all $\varphi \in C_o^\infty(\mathbb{R}^n)$. Define:

$$|||\varphi|||^2 = (P\varphi,\varphi) \; .$$

Then $|||\cdot|||$ is a Hilbert space norm on $C_o^\infty(\mathbb{R}^n)$. Completing $C_o^\infty(\mathbb{R}^n)$ in the norm $|||\cdot|||$ one obtains a Hilbert space which we denote by W. It follows from (1.7) and (1.3) that

$$W = \{u\colon u \in H^1(\mathbb{R}^n) \;,\; |V|^{1/2}u \in L^2(\mathbb{R}^n)\} \; . \tag{1.8}$$

Given $f \in L^2(\mathbb{R}^n)$ consider the map

$$\Phi_f(u) = (u,f)_{L^2} \quad \text{for} \quad u \in W$$

Then Φ_f is a bounded linear functional on W . By Riesz's representation theorem there exists a unique $v_f \in W$ such that

$$\Phi_f(u) = (u,f)_{L^2} = (u,v_f)_W \tag{1.9}$$

for all $u \in W$. Define: $T : L^2(\mathbb{R}^n) \to W \subset L^2(\mathbb{R}^n)$ by $Tf = v_f$.
Then for $f \in L^2(\mathbb{R}^n)$ with $\|f\| = 1$

$$\|Tf\|^2 \leqslant |||Tf|||^2 = (v_f,v_f)_W = (v_f,f)_{L^2} = (Tf,f)_{L^2} \leqslant \|Tf\|$$

which implies $\|Tf\| \leqslant 1$ and $(Tf,f)_{L^2} \geqslant 0$. Thus T is a self-adjoint operator with $0 \leqslant T \leqslant 1$ and $\operatorname{Ran} T \subset W$. T is also injective since if $Tf = 0$ for some $f \in L^2(\mathbb{R}^n)$, then it follows from (1.9) that $(u,f)_{L^2} = 0$ for all $u \in W$ which implies that $f = 0$.

Next it follows from the definition of Tf and the definition of W that

$$(\varphi,f)_{L^2} = (\varphi,Tf)_W = \int_{\mathbb{R}^n} (\nabla\varphi\cdot\nabla\overline{Tf} + V\varphi\overline{Tf})\,dx$$

for every $\varphi \in C_o^\infty(\mathbb{R}^n)$ and for any given $f \in L^2(\mathbb{R}^n)$. This shows that the function $u = Tf \in W$ verifies the equation $Pu = f$ in the usual weak quadratic form sense. It thus follows that $Tf \in D(H)$ and that $HTf = f$ for every $f \in L^2(\mathbb{R}^n)$. Also if $v \in D(H) \subset W$ and $Hv = f$, we have

$$(u,Hv)_{L^2} = (u,v)_W = (u,f)_{L^2}$$

for all $u \in W$ which implies that $v = Tf$. Summing up, we have shown that H is a one to one map from $D(H)$ onto $L^2(\mathbb{R}^n)$ such that $H^{-1} = T$. This proves that H is a self-adjoint positive operator.

Next observe that $D(H)$ is dense in the Hilbert space W. Indeed if this were not the case there would exist an element $v_o \in W$, $v_o \neq 0$ such that $(v_o, Tf)_W = 0$ for every $f \in L^2(\mathbb{R}^n)$. This would imply that $(v_o, f)_{L^2} = 0$ for every $f \in L^2(\mathbb{R}^n)$ (by (1.9)) leading to a contradiction.

Finally let $H_\lambda \geqslant 0$ for $\lambda \in \mathbb{R}$. Since $D(H_\lambda^{1/2})$ is the closure of $D(H)$ in the norm $(\|u\|^2 + (H_\lambda u,u))^{1/2}$ which is equivalent to the norm in W, it follows that $D(H_\lambda^{1/2}) = W$.
This yields (1.6) and completes the proof.

Given the differential operator P one can define a maximal realization $\tilde{H}$ of P in $L^2(\mathbb{R}^n)$ as follows:

$$D(\tilde{H}) = \{u: u \in L^2(\mathbb{R}^n) \cap H^1_{loc}(\mathbb{R}^n),\ Vu \in L^1_{loc}(\mathbb{R}^n),\ Pu \in L^2(\mathbb{R}^n)\}$$
$$\tilde{H}u = Pu \quad \text{for} \quad u \in D(\tilde{H}) . \tag{1.10}$$

It is clear that $\tilde{H}$ is an extension of H. In fact we have

1.2 Theorem: $\tilde{H} = H$.

1.3 Corollary: H is the only self-adjoint realization of P in $L^2(\mathbb{R}^n)$.

For the proof of the theorem we need

1.4 Lemma: Suppose that

$$(P\varphi, \varphi) \geqslant c\|\varphi\|^2 \tag{1.11}$$

for all $\varphi \in C_o^\infty(\mathbb{R}^n)$, c a positive constant. Then the map: $u \to Pu$ is an injective map from $D(\tilde{H})$ into $L^2(\mathbb{R}^n)$.

Lemma 1.4 is a corollary of Theorem 4.1 which we prove later on (see remark (iii) which follows the theorem).

Proof of Theorem 1.2: We have only to show that $D(H) \supset D(\tilde{H})$. Adding to P a constant we may assume (as in the proof of Theorem 1.1) that (1.11) holds and that H^{-1} is a well defined bounded operator in

$L^2(\mathbb{R}^n)$. Let $u \in D(\tilde{H})$ and set: $v = u - H^{-1}(\tilde{H}u)$. Then $v \in D(\tilde{H})$ and $Pv = 0$. Using Lemma 1.4 it follows that $v = 0$ which implies that $u \in D(H)$. This shows that $D(H) \supset D(\tilde{H})$ and proves the theorem.

We shall associate with the operator P the following quantities.

<u>1.5 Definition</u>:

$$\Lambda(P) = \inf\left\{\frac{(P\varphi,\varphi)}{\|\varphi\|^2} : \varphi \in C_0^\infty(\mathbb{R}^n),\ \varphi \neq 0\right\}, \tag{1.12}$$

$$\Sigma(P) = \sup_K \inf\left\{\frac{(P\varphi,\varphi)}{\|\varphi\|^2} : \varphi \in C_0^\infty(\mathbb{R}^n \setminus K),\ \varphi \neq 0\right\} \tag{1.13}$$

where the supremum is taken over all compact sets K in $\mathbb{R}^n$.

<u>1.6 Theorem</u>: Let H be the self-adjoint realization of P in $L^2(\mathbb{R}^n)$, then:

$$\inf \sigma(H) = \Lambda(P), \tag{1.14}$$

$$\inf \sigma_{ess}(H) = \Sigma(P), \tag{1.15}$$

where $\sigma(H)$ and $\sigma_{ess}(H)$ denote the spectrum and the essential spectrum of H respectively. ($\sigma_{ess}(H) = \sigma(H) \setminus \sigma_{dis}(H)$ where σ_{dis} is the discrete spectrum).

We note that formula (1.15) for the bottom of the essential spectrum was first established by Persson [5] (see also [1]).

For the proof of Theorem 1.6 as well as for results to be proved later on we need the following

<u>1.7 Lemma</u>: Suppose that

$$(P\varphi,\varphi) \geq \int_{\mathbb{R}^n \setminus K} c(x)\,|\varphi|^2 dx \tag{1.16}$$

for all $\varphi \in C_0^\infty(\mathbb{R}^n \setminus K)$ where K is a compact set in $\mathbb{R}^n$ and $c(x)$ is some real continuous function on $\mathbb{R}^n$. Then there exists a non-negative function $\chi \in C_0^\infty(\mathbb{R}^n)$ such that setting $P_\chi = P + \chi$, we have

$$(P_\chi\varphi,\varphi) \geq \int_{\mathbb{R}^n} c(x)\,|\varphi|^2 dx \tag{1.17}$$

for all $\varphi \in C_0^\infty(\mathbb{R}^n)$.

<u>Proof</u>: Choose non-negative functions: $\zeta_0 \in C_0^\infty(\mathbb{R}^n)$ and $\zeta_1 \in C^\infty(\mathbb{R}^n)$ such that $\zeta_1(x) = 0$ in a neighborhood of K and such that

$$\zeta_0^2 + \zeta_1^2 = 1 \text{ on } \mathbb{R}^n. \tag{1.18}$$

A simple computation shows that

$$\Delta = \zeta_o \Delta \zeta_o + \zeta_1 \Delta \zeta_1 + |\nabla \zeta_o|^2 + |\nabla \zeta_1|^2 \tag{1.19}$$

where the functions in (1.19) are taken as multiplication operators.

Let $\varphi \in C_o^\infty(\mathbb{R}^n)$. Using (1.18) and (1.19) we find that

$$(P\varphi,\varphi) = (P(\zeta_o\varphi),\zeta_o\varphi) + (P(\zeta_1\varphi),\zeta_1\varphi) - \int_{\mathbb{R}^n} (|\nabla\zeta_o|^2 + |\nabla\zeta_1|^2)|\varphi|^2 dx . \tag{1.20}$$

Now, it follows from (1.16) that

$$(P(\zeta_1\varphi),\zeta_1\varphi) \geq \int_{\mathbb{R}^n} c(x)\zeta_1^2|\varphi|^2 dx . \tag{1.21}$$

We also have:

$$(P(\zeta_o\varphi),\zeta_o\varphi) \geq \Lambda\int_{\mathbb{R}^n} \zeta_o^2|\varphi|^2 dx \tag{1.22}$$

where $\Lambda = \Lambda(P)$ is finite (in view of our assumptions on V). Combining (1.20), (1.21) and (1.22) it follows that

$$(P\varphi,\varphi) \geq \int_{\mathbb{R}^n} c_1(x)|\varphi|^2 dx \tag{1.23}$$

for all $\varphi \in C_o^\infty(\mathbb{R}^n)$ where

$$c_1(x) = \zeta_1^2 c(x) + \Lambda\zeta_o^2 - |\nabla\zeta_o|^2 - |\nabla\zeta_1|^2 .$$

Since $c_1(x)$ is a continuous function such that $c_1(x) = c(x)$ for $|x| \geq R_o$ for R_o sufficiently large, there exists a non-negative function $\chi \in C_o^\infty(\mathbb{R}^n)$ such that $c_1(x) + \chi(x) \geq c(x)$ on $\mathbb{R}^n$. This and (1.23) yield (1.17) and prove the lemma.

Proof of Theorem 1.6: By adding to P a constant we shall assume as before with no loss of generality that $H \geq 1$.

To prove formula (1.14) for the bottom of the spectrum one applies the spectral theorem to the operator $H^{1/2}$. Recalling that we have established before that $C_o^\infty(\mathbb{R}^n)$ is dense in $D(H^{1/2})$ (in the graph norm of $H^{1/2}$) one finds that

$$\inf \sigma(H) = \inf\left\{\frac{\|H^{1/2}u\|^2}{\|u\|^2} : u \in D(H^{1/2}),\ u \neq 0\right\}$$
$$= \inf\left\{\frac{(P\varphi,\varphi)}{\|\varphi\|^2} : \varphi \in C_o^\infty(\mathbb{R}^n),\ \varphi \neq 0\right\}$$

which proves (1.14).

We turn now to the proof of formula (1.15) for the bottom of the essential spectrum. Fix a number λ such that $\lambda < \Sigma(P)$. It follows from the definition of $\Sigma(P)$ that there exists a compact set $K \subset \mathbb{R}^n$ such that

$$(P\varphi, \varphi) \geq \lambda \|\varphi\|^2 \tag{1.24}$$

for all $\varphi \in C_o^\infty(\mathbb{R}^n \setminus K)$. An application of Lemma 1.7 to (1.24) shows that there exists a non-negative function $\chi \in C_o^\infty(\mathbb{R}^n)$ such that $((P+\chi)\varphi, \varphi) \geq \lambda\|\varphi\|^2$ for all $\varphi \in C_o^\infty(\mathbb{R}^n)$ which implies that

$$\Lambda(P_\chi) \geq \lambda \tag{1.25}$$

where we set: $P_\chi = P + \chi$. Define: $H_\chi = H + \chi$ with $D(H_\chi) = D(H)$. Then H_χ is the self-adjoint realization of P_χ in $L^2(\mathbb{R}^n)$. An application of formula (1.14) to H_χ together with (1.25) show that

$$\inf \sigma(H_\chi) = \Lambda(P_\chi) \geq \lambda . \tag{1.26}$$

Next observe that the multiplication operator $\chi : u \to \chi u$ is H-compact. (This follows from Rellich's compactness theorem [2; p. 30]). Hence by Weyl's theorem [6; p. 112] it follows that

$$\sigma_{ess}(H) = \sigma_{ess}(H_\chi) .$$

Combining this with (1.26) it follows that

$$\inf \sigma_{ess}(H) \geq \lambda$$

and since λ is an arbitrary number $< \Sigma(P)$ we conclude that

$$\inf \sigma_{ess}(H) \geq \Sigma(P) . \tag{1.27}$$

To prove the reverse inequality let μ be any positive number such that $\mu < \inf \sigma_{ess}(H)$. Let $E(\mu) = E(-\infty,\mu)$ be the spectral projection of H which corresponds to the interval $(-\infty,\mu)$. Then $E(\mu)$ is a finite rank projection so that $E(\mu) = \sum_{i=1}^{N} (\cdot,\psi_i)\psi_i$ for $\psi_i \in D(H)$. Thus if φ is a test function with support outside $B_R(0)$ we have, using Schwarz's inequality

$$\|E(\mu)\varphi\| \leq \sum_{i=1}^{N} |(\varphi, \psi_i)| \|\psi_i\|$$

$$\leq \sum_{i=1}^{N} \left\{ \int_{|x|>R} |\psi_i|^2 dx \right\}^{1/2} \|\psi_i\| \|\varphi\| .$$

Therefore, given ε, $0 < \varepsilon < 1$, there is an $R = R_\varepsilon$ so that

$$\|E(\mu)\varphi\| \leq \varepsilon\|\varphi\| \tag{1.28}$$

for all $\varphi \in C_0^\infty(\mathbb{R}^n \setminus \overline{B_R(0)})$.

Next since $C_0^\infty(\mathbb{R}^n) \subset D(H^{1/2})$ we have for any $\varphi \in C_0^\infty(\mathbb{R}^n)$:

$$\begin{aligned}(P\varphi,\varphi) &= \|H^{1/2}\varphi\|^2 \\ &= \|H^{1/2}(I-E(\mu))\varphi\|^2 + \|H^{1/2}E(\mu)\varphi\|^2 \\ &\geq \|H^{1/2}(I-E(\mu))\varphi\|^2 \geq \mu\|(I-E(\mu))\varphi\|^2 \\ &\geq \mu(\|\varphi\| - \|E(\mu)\varphi\|)^2 \quad .\end{aligned} \tag{1.29}$$

Combining (1.28) and (1.29) it follows that

$$(P\varphi,\varphi) \geq \mu(1-\varepsilon)^2\|\varphi\|^2$$

for any $\varphi \in C_0^\infty(\mathbb{R}^n \setminus \overline{B_R(0)})$, which implies that

$$\Sigma(P) \geq \mu(1-\varepsilon)^2 \quad . \tag{1.30}$$

Finally letting $\varepsilon \to 0$ in (1.30) and recalling that μ was an arbitrary positive number less than $\inf \sigma_{ess}(H)$ we get $\Sigma(P) \geq \inf \sigma_{ess}(H)$. This together with (1.27) proves (1.15) and completes the proof of the theorem.

2. Positive supersolutions ground states and bounds on eigenfunctions

In this section we shall discuss positive solutions and supersolutions of the equation:

$$(P-\lambda)u = 0 \tag{2.1}$$

in some neighborhood of infinity $\Omega_R = \{x \in \mathbb{R}^n : |x| > R\}$ and show among other things that these yield majorants for eigenfunctions of H .

We start by discussing the problem of existence of positive solutions and supersolutions of (2.1).

<u>2.1 Theorem</u>: Under the same conditions on P as before, suppose that for some real λ the equation (2.1) admits in some neighborhood of infinity Ω_R a positive and continuous supersolution, then

$$\lambda \leq \Sigma(P) = \inf \sigma_{ess}(H) . \tag{2.2}$$

If (2.1) admits a positive continuous supersolution in $\mathbb{R}^n$, then

$$\lambda \leq \Lambda(P) = \inf \sigma(H) . \tag{2.3}$$

Proof: Suppose there exists a positive function $v \in C(\Omega_R) \cap H^1_{loc}(\Omega_R)$ such that

$$(P-\lambda)v \geq 0 \tag{2.4}$$

in Ω_R. Let $\varphi \in C_o^\infty(\Omega_R)$, φ real. Noting that φ^2/v is a non-negative function in $C_o(\Omega_R) \cap H^1(\Omega_R)$, it follows from (2.4) that

$$\int_{\Omega_R} [\nabla v \cdot \nabla(\varphi^2/v) + (V-\lambda)v(\varphi^2/v)]dx \geq 0 . \tag{2.5}$$

We shall use the identity

$$\nabla v \cdot \nabla(\varphi^2/v) = |\nabla\varphi|^2 - v^2|\nabla(\varphi/v)|^2 . \tag{2.6}$$

From (2.5) and (2.6) it follows that

$$((P-\lambda)\varphi,\varphi) \geq \int_{\Omega_R} v^2|\nabla(\varphi/v)|^2 dx \geq 0 . \tag{2.7}$$

Although φ was assumed to be real it is clear that (2.7) also holds for all complex $\varphi \in C_o^\infty(\Omega_R)$. Invoking Definition 1.5 and Theorem 1.6 it follows from (2.7) that (2.2) holds.

In case there exists a positive and continuous supersolution v of (2.1) in $\mathbb{R}^n$ one finds in the same way that $((P-\lambda)\varphi,\varphi) \geq 0$ for all $\varphi \in C_o^\infty(\mathbb{R}^n)$ which implies (2.3). This completes the proof.

In the remainder of this section we shall assume in addition to the conditions imposed before that $V \in L^p_{loc}(\mathbb{R}^n)$ for some $p > n/2$. By Theorem 0.1 this condition ensures that all solutions of (2.1) in a connected open set Ω are continuous and that a non-negative solution of (2.1) in Ω cannot vanish at a point without vanishing identically.

2.2 Theorem: Let H be the self-adjoint realization of P. Suppose that $\Lambda = \inf \sigma(H)$ is an eigenvalue of H. Then Λ is a simple eigenvalue. Furthermore, up to a multiplicative constant the eigenfunction ψ which corresponds to Λ is a positive continuous function.

2.3 Definition: The positive eigenfunction ψ which corresponds to Λ (when it exists) is called a ground state of H.

2.4 Corollary: It follows from Theorem 2.2 that if $\inf \sigma(H) < \inf \sigma_{ess}(H)$ then H has a ground state.

2.5 Corollary: It follows from Theorem 2.1 and Theorem 2.2 that if ψ is a positive eigenfunction of H with eigenvalue λ then

$\lambda = \inf \sigma(H)$ and ψ is a ground state.

Proof of Theorem 2.2: We assume with no loss of generality that $\Lambda = 0$.

Let ψ be a non-null function in $D(H)$ such that $H\psi = 0$. It follows from our assumptions on V that $\psi(x)$ is a continuous function. Consider the function $|\psi|$. Since $\psi \in H^1(\mathbb{R}^n)$ it follows from a well known result (see [4 ; § 7.4]) that $|\psi| \in H^1(\mathbb{R}^n)$ and that

$$|\nabla|\psi(x)||^2 = |\nabla\psi(x)|^2 \tag{2.8}$$

a.e. in $\mathbb{R}^n$. Noting that $H \geqslant 0$, it follows from Theorem 1.1 that $|\psi| \in D(H^{1/2})$. Using (2.8) we also find that

$$(P|\psi|,|\psi|) = (P\psi,\psi) = 0 . \tag{2.9}$$

Since $(Pv,v) \geqslant 0$ for all $v \in D(H^{1/2})$, it follows from (2.9) that

$$(P|\psi|,\varphi) = 0 \tag{2.10}$$

for all $\varphi \in C_o^\infty(\mathbb{R}^n) \subset D(H^{1/2})$. This proves that $|\psi| \in D(H)$ and that $P|\psi| = 0$. By Harnack's inequality applied to $|\psi|$ it follows further that $\psi(x) \neq 0$ for all x.

The above considerations show that if ψ is any real eigenfunction of H with eigenvalue Λ then either ψ or $-\psi$ is a strictly positive function. This clearly implies the theorem and completes the proof.

The following theorem is a kind of converse of Theorem 2.1.

2.6 Theorem: Let P be a Schrödinger differential operator and let H be its realization in $L^2(\mathbb{R}^n)$ as above. For $\lambda \in \mathbb{R}$ we have:

(i) If $\lambda < \inf \sigma_{ess}(H)$ then there exists a positive solution of the equation $(P-\lambda)u = 0$ in some neighborhood of infinity in $\mathbb{R}^n$.

(ii) If $\lambda \leqslant \inf \sigma(H)$ then there exists a positive solution of the equation $(P-\lambda)u = 0$ in $\mathbb{R}^n$.

Proof: We shall prove (i) and sketch the proof of (ii).

To prove (i) fix a number μ such that $\lambda < \mu < \inf \sigma_{ess}(H)$. Applying Lemma 1.7 and Persson's formula (1.15) it follows that there exists a non-negative function $\chi \in C_o^\infty(\mathbb{R}^n)$ such that

$$\inf \sigma(H+\chi) \geqslant \mu . \tag{2.11}$$

Consider the family of operators $H_t = H + t\chi$, $t \in \mathbb{R}$. Set

$$\Lambda(t) = \inf \sigma(H_t) . \tag{2.12}$$

Using (1.12) it follows that $\Lambda(t)$ is a continuous function of t such that $\Lambda(1) \geq \mu$, $\Lambda(t) \to -\infty$ as $t \to -\infty$. By continuity there exists a number $t_o < 1$ such that $\Lambda(t_o) = \lambda$. Hence we find that

$$\inf \sigma(H_{t_o}) = \lambda ,$$

$$\inf \sigma_{ess}(H_{t_o}) = \inf \sigma_{ess}(H) > \lambda$$

where the last relation follows from Weyl's theorem. Applying Corollary 2.4 it follows that H_{t_o} has a ground state ψ with an eigenvalue λ . Cleary ψ is a positive solution of $(P-\lambda)u = 0$ in $\mathbb{R}^n \setminus \operatorname{supp} \chi$. This proves (i).

We shall sketch the proof of (ii). Suppose first that $\lambda < \inf \sigma(H)$. Let χ be <u>any</u> non-negative function in $C_o^\infty(\mathbb{R}^n)$, $\chi \not\equiv 0$. The same argument used in the proof of (i) shows that there exists a number t_o such that $H + t_o\chi$ has a ground state with an eigenvalue λ . This shows that the equation $(P-\lambda)u = 0$ admits a positive solution in $\mathbb{R}^n \setminus \bar{B}$ where B is <u>any</u> given ball in $\mathbb{R}^n$. Let now $\{x^j\}$, $j = 1,2,\ldots$, be a sequence of points in $\mathbb{R}^n$ such that $|x^j| > 1$, $|x^j| \to \infty$ as $j \to \infty$. By the last <u>remark</u> there exists a positive solution u_j of $(P-\lambda)u = 0$ in $\mathbb{R}^n \setminus B_1(x^j)$ with $u_j(0) = 1$, $j = 1,2,\ldots$ Applying Harnack's inequality it follows that there exists a subsequence $\{u_{j_k}\}$ which converges uniformly on every compact set to a positive function u in $\mathbb{R}^n$. By "standard" elliptic estimates it follows further that $u \in H^1_{loc}(\mathbb{R}^n)$ and that $(P-\lambda)u = 0$ in $\mathbb{R}^n$. This gives (ii) when $\lambda < \inf \sigma(H)$.

Finally to prove (ii) for $\lambda = \inf \sigma(H)$ one notes that by the result just proved there exists a sequence of positive functions $\{v_j\}$ with $v_j(0) = 1$, such that v_j is a solution of the equation $(P-\lambda+j^{-1})v = 0$ in $\mathbb{R}^n$. Using again Harnack's inequality one shows that a subsequence of $\{v_j\}$ converges to a positive solution of the equation $(P-\lambda)u = 0$ in $\mathbb{R}^n$, thus completing the proof of (ii).

We conclude this section by showing that positive supersolutions furnish majorants for eigenfunctions of H . This follows from

<u>2.7 Theorem</u>: Let w be a positive and continuous supersolution of the equation $(P-\lambda)u = 0$ in a neighborhood of infinity Ω_R . Let v be a continuous subsolution of the same equation in Ω_R . Suppose that

$$\liminf_{N\to\infty}\left(N^{-2}\int_{N<|x|<\alpha N}|v|^2dx\right) = 0 \tag{2.13}$$

for some number $\alpha > 1$. Then there exists a constant C such that

$$v(x) \leqslant Cw(x) \tag{2.14}$$

in Ω_{R+1} .

<u>2.8 Corollary</u>: Let ψ be an eigenfunction of H with eigenvalue λ . Let w be a positive and continuous supersolution of the equation $(P-\lambda)u = 0$ in Ω_R . Then there exists a constant C such that

$$|\psi(x)| \leqslant Cw(x) \tag{2.15}$$

in Ω_{R+1} .

For the proof of Theorem 2.7 we need

<u>2.9 Lemma</u>: Let v be a subsolution of the equation $(P-\lambda)u = 0$ in an open set Ω . Then $v_+ = \max(v,0)$ is also a subsolution of the same equation in Ω .

Under certain assumptions on V this Lemma is well known. The following proof shows that the Lemma is valid under the sole condition that $V \in L^1_{loc}(\mathbb{R}^n)$.

<u>Proof of Lemma</u>: By assumption $v \in H^1_{loc}(\Omega)$ and $Vu \in L^1_{loc}(\Omega)$. For any $\varepsilon > 0$ define: $v_\varepsilon = (v^2+\varepsilon^2)^{1/2}$. By well known results it follows that the functions v_ε , $|v|$ and v_+ are in $H^1_{loc}(\Omega)$ (see for example [4 ; § 7.4]). It is also easy to see that

$$v_\varepsilon \to |v| \text{ in } H^1_{loc}(\Omega) \quad \text{as} \quad \varepsilon \to 0 \quad . \tag{2.16}$$

Now let φ be a non-negative function in $C_o^\infty(\mathbb{R}^n)$. A simple computation shows that

$$\nabla v_\varepsilon \cdot \nabla\varphi = \nabla v \cdot \nabla\left(\frac{v}{v_\varepsilon}\varphi\right) - \varphi v_\varepsilon^{-3}(v_\varepsilon^2 - v^2)\,|\nabla v|^2$$

a.e. in Ω , which implies that

$$\nabla v_\varepsilon \cdot \nabla\varphi \leqslant \nabla v \cdot \nabla\left(\frac{v}{v_\varepsilon}\varphi\right) . \tag{2.17}$$

Set $\varphi_\varepsilon = \frac{1}{2}\left(1+\frac{v}{v_\varepsilon}\right)\varphi$. It follows from (2.17) that

$$\nabla \tfrac{1}{2}(v_\varepsilon + v)\cdot\nabla\varphi \leqslant \nabla v \cdot \nabla\varphi_\varepsilon \quad . \tag{2.18}$$

Noting that φ_ε is a non-negative function with compact support in Ω and that $\varphi_\varepsilon \in H^1(\Omega) \cap L^\infty(\Omega)$, we use φ_ε as a test function against the subsolution v. We have:

$$\int_\Omega [\nabla v\cdot\nabla\varphi_\varepsilon + (V-\lambda)v\varphi_\varepsilon]dx \leq 0 ,$$

which when combined with (2.18) yields

$$\int_\Omega \nabla\frac{1}{2}(v_\varepsilon+v)\cdot\nabla\varphi\, dx \leq -\int_\Omega (V-\lambda)v\varphi_\varepsilon dx \quad . \tag{2.19}$$

Finally, letting $\varepsilon \to 0$, using (2.16) and the fact that $0 \leq \varphi_\varepsilon \leq \varphi$, $\varphi_\varepsilon \to (\mathrm{sgn}\, v_+)\varphi$ as $\varepsilon \to 0$, it follows from (2.19) that

$$\int_\Omega [\nabla v_+\cdot\nabla\varphi + (V-\lambda)v_+\varphi]dx \leq 0$$

for any non-negative $\varphi \in C_0^\infty(\Omega)$. This proves the Lemma.

<u>Proof of Theorem 2.7</u>: Set $R_0 = R+1$ and choose a constant $C>0$ such that

$$Cw(x) - v(x) > 0 \quad \text{on the sphere:} \quad |x| = R_0 . \tag{2.20}$$

Define a function u_0 in Ω_{R_0} by

$$u_0(x) = (v(x)-Cw(x))_+ . \tag{2.21}$$

We shall prove the theorem by showing that $u_0(x) \equiv 0$ in Ω_{R_0}.

We start by observing that u_0 is a continuous subsolution of the equation $(P-\lambda)u = 0$ in Ω_{R_0}. This follows from Lemma 2.9.

Since $w>0$ it follows from (2.21) that $0 \leq u_0 \leq v_+$ which in view of (2.13) implies that

$$\liminf_{N\to\infty}\left(N^{-2}\int_{N<|x|<\alpha N} u_0^2 dx\right) = 0 \tag{2.22}$$

Also, it follows from (2.20), (2.21) and the continuity of v and w that

$$u_0(x) = 0 \text{ for } R_0 < |x| < R_0 + \delta \tag{2.23}$$

for some $\delta > 0$.

Since u_0 is a non-negative subsolution we have:

$$\int_{\Omega_{R_0}} [\nabla u_0\cdot\nabla(\zeta^2 u_0) + (V-\lambda)u_0(\zeta^2 u_0)]dx \leq 0 \tag{2.24}$$

for any real function $\zeta \in C_0^\infty(\Omega_{R_0})$. Moreover, in view of (2.23) it follows that (2.24) holds for any real function $\zeta \in C_0^\infty(\mathbb{R}^n)$. Using the identity:

$$\nabla u_0\cdot\nabla(\zeta^2 u_0) = |\nabla(\zeta u_0)|^2 - u_0^2|\nabla\zeta|^2 \tag{2.25}$$

it follows from (2.24) that

$$\int_{\Omega_{R_o}} [|\nabla(\zeta u_o)|^2 + (V-\lambda)(\zeta u_o)^2]dx \leq \int_{\Omega_{R_o}} |\nabla\zeta|^2 u_o^2 dx \tag{2.26}$$

for any real $\zeta \in C_o^\infty(\mathbb{R}^n)$. Next we use the fact that there exists a positive and continuous supersolution w of the equation $(P-\lambda)u = 0$ in $\Omega_R \supset \bar{\Omega}_{R_o}$. By the same argument leading to the inequality (2.7) we find that

$$\int_{\Omega_{R_o}} w^2 \left|\nabla\left(\frac{\zeta u_o}{w}\right)\right|^2 dx \leq \int_{\Omega_{R_o}} [|\nabla(\zeta u_o)|^2 + (V-\lambda)(\zeta u_o)^2]dx . \tag{2.27}$$

Combining (2.27) with (2.26) we find that

$$\int_{\Omega_{R_o}} w^2 \left|\nabla\left(\frac{\zeta u_o}{w}\right)\right|^2 dx \leq \int_{\Omega_{R_o}} |\nabla\zeta|^2 u_o^2 dx \tag{2.28}$$

for any real $\zeta \in C_o^\infty(\mathbb{R}^n)$.

We now pick a function $\chi \in C_o^\infty(\mathbb{R}^n)$ such that $0 \leq \chi \leq 1$, $\chi(x) = 1$ for $|x| \leq 1$, $\chi(x) = 0$ for $|x| \geq \alpha$, where $\alpha > 1$ is the number in (2.22). We set for $N \geq 1$: $\chi_N(x) = \chi(x/N)$. Applying (2.28) with $\zeta = \chi_N$, letting $N \to \infty$, using (2.22) and Fatou's Lemma, we find that

$$\int_{\Omega_{R_o}} w^2 \left|\nabla\left(\frac{u_o}{w}\right)\right|^2 dx \leq \liminf_{N\to\infty} \int_{\Omega_{R_o}} |\nabla\chi_N|^2 u_o^2 dx$$

$$\leq \text{Const.} \liminf_{N\to\infty} \left(N^{-2} \int_{N<|x|<\alpha N} u_o^2 dx\right) = 0 ,$$

which implies that $u_o = cw$ in Ω_{R_o} where c is some constant. Since $w > 0$ in Ω_{R_o} while u_o vanishes in some boundary strip (by (2.23)) it follows that $c = 0$. Thus $u_o(x) \equiv 0$ in Ω_{R_o}. This proves the theorem.

3. Upper and lower bounds on eigenfunctions of two body Schrödinger operators

We shall say that $P = -\Delta + V$ is a two body Schrödinger operator if V is a real function in $L^p_{loc}(\mathbb{R}^n)$ with $p > n/2$ such that $V(x) \to 0$ as $|x| \to \infty$. If H is the self-adjoint realization of a two body Schrödinger operator then it is easy to see that $\sigma_{ess}(H) = [0,\infty)$. (By application of formula (1.15) it follows that $\inf \sigma_{ess}(H) = 0$.)

In this section we shall derive upper and lower bounds on eigenfunctions of certain classes of two body Schrödinger operators. This will be done by constructing positive supersolutions and subsolutions of $(P-\lambda)u = 0$ for $\lambda < 0$ in some neighborhood of infinity and applying Theorem 2.7.

3.1 Definition: A two body potential $V(x)$ is said to be short range if

$$V(x) = O(|x|^{-1-\delta}) \quad \text{as} \quad |x| \to \infty \tag{3.1}$$

for some $\delta > 0$.

3.2 Theorem: Let $P = -\Delta + V$ be a Schrödinger differential operator with a short range potential V. Let u be a solution of the differential equation $(P-\lambda)u = 0$ in $\Omega_R = \{x: |x| > R\}$, $\lambda < 0$. Set $k = |\lambda|^{1/2}$. We have:

(i) If $u \in L^2(\Omega_R)$, then

$$|u(x)| \leq C|x|^{-(n-1)/2} e^{-k|x|} \quad \text{in} \quad \Omega_{R+1}. \tag{3.2}$$

(ii) If u is positive in Ω_R (u not necessarily in $L^2(\Omega_R)$), then

$$u(x) \geq c|x|^{-(n-1)/2} e^{-k|x|} \quad \text{in} \quad \Omega_{R+1}. \tag{3.3}$$

Here C and c are positive constants.

It follows from Theorem 3.2 that if H is the self-adjoint realization of a Schrödinger differential operator with a short range potential and if $\Lambda = \inf \sigma(H) < 0$, then the ground state ψ of H satisfies the upper and lower bounds

$$C_1|x|^{-(n-1)/2} e^{-k|x|} \leq \psi(x) \leq C_2|x|^{-(n-1)/2} e^{-k|x|} \tag{3.4}$$

for $|x| \geq 1$ where $k = |\Lambda|^{1/2}$ and C_i are certain positive constants.

The estimates (3.2) and (3.3) need not hold if V is a two body potential which is not short range. (For instance if $V = -|x|^{-1}$ then the ground state ψ does not verify the upper bound (3.4).) We shall establish the following result which contains Theorem 3.2 as a special case and which applies to a class of two body Schrödinger operators with long range potentials.

3.3 Theorem: Let $P = -\Delta + V$ be a two body Schrödinger operator. Suppose that V has a decomposition:

$$V(x) = V_S(x) + W(x) \tag{3.5}$$

where V_S is a short range potential in $\mathbb{R}^n$ and W is a real function in $C^2(\mathbb{R}^n)$ such that as $|x| \to \infty$ the following relations hold:

$$W(x) = O(|x|^{-1/2-\delta}),$$
$$\partial W(x) = O(|x|^{-3/2-\delta}), \quad \partial^2 W(x) = O(|x|^{-2-\delta}) \tag{3.6}$$

for some $0<\delta<1/2$, where ∂^ℓ stands for any partial derivative of order ℓ . Define:

$$F(x) = \int_0^{|x|} W(\frac{x}{|x|}t)\,dt . \tag{3.7}$$

Let u be a solution of the differential equation

$$(P-\lambda)u = 0 \tag{3.8}$$

in Ω_R , $\lambda<0$. Set $k = |\lambda|^{1/2}$. We have

(i) If $u \in L^2(\Omega_k)$, then

$$|u(x)| \leqslant C|x|^{-(n-1)/2} e^{-k|x|-(2k)^{-1}F(x)} \quad \text{in } \Omega_{R+1} . \tag{3.9}$$

(ii) If u is positive in Ω_R , then

$$u(x) \geqslant c|x|^{-(n-1)/2} e^{-k|x|-(2k)^{-1}F(x)} \quad \text{in } \Omega_{R+1} . \tag{3.10}$$

Here C and c are certain positive constants.

<u>Example</u>: Suppose that $V(x) = \gamma|x|^{-1}$ + a short range potential, γ a constant. Then the ground state ψ of $-\Delta+V$ (when $\inf\sigma(H) = -k^2$, $k>0$) satisfies:

$$C_1|x|^{-\alpha}e^{-k|x|} \leqslant \psi(x) \leqslant C_2|x|^{-\alpha}e^{-k|x|}$$

for $|x|\geqslant 1$ where $\alpha = (n-1)/2 + \gamma/(2k)$.

<u>Proof of Theorem 3.3</u>: Replacing δ if necessary by a smaller positive number we shall assume with no loss of generality that

$$V_S(x) = O(|x|^{-1-\delta}) \quad \text{as} \quad |x| \to \infty . \tag{3.11}$$

Next observe that $F \in C^2(R^n\setminus\{0\})$. Moreover, by a straightforward computation (which we shall omit) it follows from (3.6) and (3.7) that

$$F(x) = O(|x|^{1/2-\delta}) ,$$

$$\partial F(x) = O(|x|^{-1/2-\delta}) ,$$

and

$$\partial^2 F(x) = O(|x|^{-1-\delta}) , \tag{3.12}$$

as $|x| \to \infty$.

Define:

$$\Phi^+(x) = k|x| + (2k)^{-1}F(x) + |x|^{-\varepsilon} , \tag{3.13}$$

$$\Phi^-(x) = k|x| + (2k)^{-1}F(x) - |x|^{-\varepsilon} ,$$

where $0<\varepsilon<\delta$ and set:

$$w(x) = |x|^{-(n-1)/2} e^{-\Phi^+(x)} \tag{3.14}$$

$$v(x) = |x|^{-(n-1)/2} e^{-\Phi^-(x)} \quad .$$

We claim that $w(x)$ is a supersolution of (3.8) in Ω_{R_o} and that $v(x)$ is a subsolution of (3.8) in Ω_{R_o} for some $R_o > 0$. Assuming for a moment the validity of our claim we derive the two results of the theorem.

Suppose first that u is a solution of (3.8) in Ω_R such that $u \in L^2(\Omega_R)$. Applying Theorem 2.7 to the subsolution $|u|$ (invoking Lemma 2.9) and the supersolution w of (3.8) in Ω_{R_1}, $R_1 = \max(R, R_o)$, it follows that

$$|u(x)| \leqslant C_1 |x|^{-(n-1)/2} e^{-\Phi^+(x)}$$

$$\leqslant C_2 |x|^{-(n-1)/2} e^{-k|x| - (2k)^{-1} F(x)}$$

for $|x| > R_1 + 1$ which yields (3.9).

Suppose next that u is a positive solution of (3.8) in Ω_R. Applying Theorem 2.7 to the subsolution v and positive solution u in Ω_{R_1} (note that $v \in L^2(\Omega_{R_1})$) we get

$$|x|^{-(n-1)/2} e^{-\Phi^-(x)} \leqslant C_3 u(x)$$

for $|x| > R_1 + 1$ which in view of (3.13) yields the lower bound (3.10).

We conclude the proof by showing that w is a supersolution of (3.8) in some neighborhood of infinity in R^n (the proof that v is a subsolution is similar and shall be omitted).

Now a simple computation shows that

$$\Delta\left(r^{-\frac{n-1}{2}} e^{-\Phi^+}\right) = r^{-\frac{n-1}{2}} e^{-\Phi^+}\left(|\nabla\Phi^+|^2 - \Delta\Phi^+ + \frac{n-1}{r}\frac{\partial\Phi^+}{\partial r} - \frac{(n-1)(n-3)}{4r^2}\right) \tag{3.15}$$

where $r = |x|$. Using (3.13), we find that

$$\nabla\Phi^+ = k\frac{x}{|x|} + (2k)^{-1}\nabla F - \varepsilon|x|^{-\varepsilon-1}\frac{x}{|x|} \quad ,$$

which in view of (3.7) and (3.12) gives:

$$|\nabla\Phi^+|^2 = k^2 + \frac{\partial F}{\partial r} - 2k\varepsilon r^{-1-\varepsilon} + O(r^{-1-\delta})$$

$$= k^2 + W(x) - 2k\varepsilon r^{-1-\varepsilon} + O(r^{-1-\delta}) \tag{3.16}$$

as $r \to \infty$. Using (3.13) and (3.6) we find that

$$\Delta\Phi^+ - \frac{n-1}{r}\frac{\partial\Phi^+}{\partial r} = \left(\Delta - \frac{n-1}{r}\frac{\partial}{\partial r}\right)\left((2k)^{-1}F + r^{-\varepsilon}\right)$$
$$= O(r^{-1-\delta}) \quad . \tag{3.17}$$

Hence using (3.14), (3.15), (3.16), (3.17) and (3.11) we find that

$$w^{-1}(P-\lambda)w = w^{-1}(-\Delta+V+k^2)w =$$
$$= (-|\nabla\Phi^+|^2 + \Delta\Phi^+ - \frac{n-1}{r}\frac{\partial\Phi^+}{\partial r} + \frac{(n-1)(n-3)}{4r^2} + V_S + W+k^2)$$
$$= -k^2 - W + 2k\varepsilon r^{-1-\varepsilon} + W + k^2 + O(r^{-1-\delta})$$
$$= 2k\varepsilon r^{-1-\varepsilon} + O(r^{-1-\delta}) \quad \text{as} \quad r \to \infty \quad ,$$
$$> 0 \quad \text{for} \quad r > R_o \quad \text{sufficiently large since} \quad 0 < \varepsilon < \delta \ .$$

Hence w is a supersolution in a neighborhood of infinity as claimed.

4. Weighted L^2 estimates

The comparison method used in the previous section to derive bounds on eigenfunctions of 2-body Schrödinger operators does not seem to be an effective method to derive upperbounds on eigenfunctions of N-body Schrödinger operators when $N \geq 3$. In this section we shall introduce another method based on L^2 weighted estimates which yields upperbounds on eigenfunctions of Schrödinger operators in a general situation. The method will be applied later to the (generalized) N-body Schrödinger operator.

We consider as before a Schrödinger operator $P = -\Delta+V$ on $\mathbb{R}^n$. Unless otherwise stated we shall assume that V satisfies the same conditions as in section 1 (namely V is a real function in $L^1_{loc}(\mathbb{R}^n)$ such that V_- satisfies (1.2)). The key weighted L^2 estimate is given in the following

4.1 Theorem: Suppose that for some $\lambda \in \mathbb{R}$ and some positive continuous function $c(x)$ on $\mathbb{R}^n$ we have: $P-\lambda \geq c(\cdot)$ in the sense that

$$\int_{\mathbb{R}^n} (|\nabla\varphi|^2 + (V-\lambda)|\varphi|^2)\,dx \geq \int_{\mathbb{R}^n} c(x)|\varphi|^2 dx \tag{4.1}$$

for every $\varphi \in C_o^\infty(\mathbb{R}^n)$.

Let u be a solution of the equation

$$(P-\lambda)u = f \tag{4.2}$$

in $\mathbb{R}^n$ where $f \in L^2_{loc}(\mathbb{R}^n)$. Suppose that u verifies the growth condition

$$\liminf_{R\to\infty}\left(R^{-2}\int_{R<|x|<\alpha R}|u|^2dx\right) = 0 \tag{4.3}$$

for some fixed number $\alpha > 1$. Let $h(x)$ be a non-negative Lipschitz function on $\mathbb{R}^n$ such that

$$|\nabla h(x)|^2 < c(x) \tag{4.4}$$

a.e. in $\mathbb{R}^n$. Then the following inequality holds:

$$\int_{\mathbb{R}^n}(c(x) - |\nabla h(x)|^2)e^{2h(x)}|u|^2dx \leq \int_{\mathbb{R}^n}(c(x)-|\nabla h(x)|^2)^{-1}e^{2h(x)}|f|^2dx . \tag{4.5}$$

Remarks:

i) In connection with (4.4) we recall that by a theorem of Rademacher a Lipschitz function has a differential almost everywhere.

ii) The growth restriction (4.3) can be replaced by other growth restrictions (see [1 ; Theorem 1.5]).

iii) A simple corollary of Theorem 4.1 is that if u is a solution of the equation $(P-\lambda)u = 0$ in $\mathbb{R}^n$ and if condition (4.1) holds then the assumption $u \in L^2(\mathbb{R}^n)$ implies that u is a null function. (Apply the theorem with $h \equiv 0$.) This in particular proves Lemma 1.4 where it is assumed that (4.1) holds (for $\lambda = 0$) with $c(x)$ a positive constant.

Proof of Theorem 4.1: We shall assume that the right hand side of (4.5) is finite since otherwise the inequality to be proved is trivial. Since u is a solution of (4.2) we have (by definition) that $u \in H^1_{loc}(\mathbb{R}^n)$, $Vu \in L^1_{loc}(\mathbb{R}^n)$, and that

$$\int_{\mathbb{R}^n}(\nabla u\cdot\nabla\varphi+(V-\lambda)u\varphi)\,dx = \int_{\mathbb{R}^n}f\varphi\,dx \tag{4.6}$$

for every $\varphi \in C^\infty_o(\mathbb{R}^n)$. To prove the inequality (4.5) we shall assume with no loss of generality that u and f are real.

For any $\varepsilon > 0$ define: $u_\varepsilon = u/(1+\varepsilon u^2)$. Then $u_\varepsilon \in L^\infty(\mathbb{R}^n)$ and a standard approximation argument shows that $u_\varepsilon \in H^1_{loc}(\mathbb{R}^n)$ and that $u_\varepsilon \to u$ in $H^1_{loc}(\mathbb{R}^n)$ as $\varepsilon \to 0$. Let now ψ be a real Lipschitz function of compact support. By another approximation argument it follows that $u_\varepsilon\psi^2 \in H^1_{loc}(\mathbb{R}^n) \cap L^\infty_o(\mathbb{R}^n)$ and that (4.6) holds with $\varphi = u_\varepsilon\psi^2$. Thus we obtain that

$$\int_{\mathbb{R}^n}(\nabla u\cdot\nabla(u_\varepsilon\psi^2) + (V-\lambda)uu_\varepsilon\psi^2)\,dx = \int_{\mathbb{R}^n}fu_\varepsilon\psi^2dx \tag{4.7}$$

We rewrite (4.7) in the form

$$\int_{\mathbb{R}^n} (\nabla u_\varepsilon \cdot \nabla(u_\varepsilon \psi^2) + (V-\lambda) u_\varepsilon^2 \psi^2)\, dx = \int_{\mathbb{R}^n} f u_\varepsilon \psi^2 dx + I_\varepsilon \tag{4.8}$$

where

$$I_\varepsilon = \int_{\mathbb{R}^n} (\nabla(u_\varepsilon - u) \cdot \nabla(u_\varepsilon \psi^2) + (V-\lambda)(u_\varepsilon - u) u_\varepsilon \psi^2)\, dx$$

$$\leqslant \int_{\mathrm{supp}\,\psi} (\nabla(u_\varepsilon - u) \cdot \nabla(u_\varepsilon \psi^2) + (V_- + \lambda)(u - u_\varepsilon) u_\varepsilon \psi^2)\, dx . \tag{4.9}$$

Since $u_\varepsilon \to u$ in $H^1_{loc}(\mathbb{R}^n)$, and in view of our assumption (1.2) on V_-, it follows readily that the last integral in (4.9) tends to zero as $\varepsilon \to 0$, so that we have:

$$\limsup_{\varepsilon \to 0} I_\varepsilon \leqslant 0 . \tag{4.10}$$

Using the identity: $\nabla u_\varepsilon \cdot \nabla(u_\varepsilon \psi^2) = |\nabla(u_\varepsilon \psi)|^2 - u_\varepsilon^2 |\nabla \psi|^2$, we now rewrite (4.8) in the form

$$\int_{\mathbb{R}^n} (|\nabla(u_\varepsilon \psi)|^2 + (V-\lambda) u_\varepsilon^2 \psi^2 - u_\varepsilon^2 |\nabla\psi|^2)\, dx = \int_{\mathbb{R}^n} f u_\varepsilon \psi^2 dx + I_\varepsilon . \tag{4.11}$$

From our assumption (4.1) and a density argument it follows that

$$\int_{\mathbb{R}^n} (|\nabla(u_\varepsilon \psi)|^2 + (V-\lambda)(u_\varepsilon \psi)^2)\, dx \geqslant \int_{\mathbb{R}^n} c(x)(u_\varepsilon \psi)^2 dx . \tag{4.12}$$

Hence combining (4.11) and (4.12) we get

$$\int_{\mathbb{R}^n} (c(x)(u_\varepsilon \psi)^2 - |\nabla\psi|^2 u_\varepsilon^2)\, dx \leqslant \int_{\mathbb{R}^n} f u_\varepsilon \psi^2 dx + I_\varepsilon . \tag{4.13}$$

Letting $\varepsilon \to 0$ and using (4.10) it follows from (4.13) that

$$\int_{\mathbb{R}^n} (c(x)(u\psi)^2 - |\nabla\psi|^2 u^2)\, dx \leqslant \int_{\mathbb{R}^n} f u \psi^2 dx . \tag{4.14}$$

Let $h(x)$ be the non-negative Lipschitz function introduced in the theorem. Define:

$$h_N(x) = \frac{N-1}{N} h(x) \quad \text{if} \quad h(x) \leqslant N ,$$

$$h_N(x) = N-1 \quad \text{if} \quad h(x) > N , \tag{4.15}$$

$N = 1,2,\ldots$. It is clear that $h_N(x)$ is a bounded non-negative Lipschitz function on $\mathbb{R}^n$. Moreover, using Rademacher's theorem, (4.15) and (4.4), it follows readily that

$$|\nabla h_N(x)|^2 \leqslant \left(\frac{N-1}{N}\right)^2 |\nabla h(x)|^2 < \left(\frac{N-1}{N}\right)^2 c(x) . \tag{4.16}$$

Next we pick a function $\chi \in C_o^\infty(\mathbb{R}^n)$ such that $0 \leq \chi \leq 1$ on $\mathbb{R}^n$ and such that $\chi(x) = 1$ for $|x| \leq 1$, $\chi(x) = 0$ for $|x| \geq \alpha$ where $\alpha > 1$ is the number in (4.3). We set:

$$\chi_R(x) = \chi(x/R) \quad \text{for} \quad R \geq 1 ,$$

and let

$$\psi_{R,N}(x) = \chi_R(x)\, e^{h_N(x)} . \tag{4.17}$$

Note that

$$\begin{aligned} |\nabla\psi_{R,N}|^2 &= (\chi_R^2|\nabla h_N|^2 + 2\chi_R\nabla\chi_R\cdot\nabla h_N + |\nabla\chi_R|^2)\, e^{2h_N} \\ &\leq (\chi_R^2|\nabla h_N|^2(1+\tfrac{1}{N}) + (1+N)|\nabla\chi_R|^2)\, e^{2h_N} \\ &\leq \psi_{R,N}(x)^2|\nabla h|^2 + (1+N)|\nabla\chi_R|^2\, e^{2h_N} \end{aligned} \tag{4.18}$$

where we have used (4.16).

We now apply the inequality (4.14) with $\psi = \psi_{R,N}$. Using (4.18), we obtain

$$\begin{aligned} \int (u\psi_{R,N})^2(c(x)-|\nabla h(x)|^2)\,dx &\leq \int_{\mathbb{R}^n} fu\psi_{R,N}^2 dx \\ &\quad + (1+N)\int_{\mathbb{R}^n} u^2|\nabla\chi_R|^2 e^{2h_N} dx \\ &\leq \Big(\int_{\mathbb{R}^n} (u\psi_{R,N})^2(c-|\nabla h|^2)\,dx\Big)^{1/2}\Big(\int_{\mathbb{R}^n}(f\psi_{R,N})^2(c-|\nabla h|^2)^{-1}dx\Big)^{1/2} \\ &\quad + (1+N)\int_{\mathbb{R}^n} u^2|\nabla\chi_R|^2 e^{2h_N} dx , \end{aligned} \tag{4.19}$$

where Schwarz's inequality was used to derive the last inequality. By an elementary calculation it follows from (4.19) that

$$\begin{aligned} \int_{\mathbb{R}^n} (u\psi_{R,N})^2(c(x)-|\nabla h(x)|^2)\,dx &\leq \int_{\mathbb{R}^n}(f\psi_{R,N})^2(c(x)-|\nabla h(x)|^2)^{-1}dx \\ &\quad + 2(1+N)\int_{\mathbb{R}^n} u^2|\nabla\chi_R|^2 e^{2h_N} dx . \end{aligned} \tag{4.20}$$

Noting that $|\nabla\chi_R(x)| = O(R^{-1})$ as $R \to \infty$ uniformly in x, and also that $\operatorname{supp}|\nabla\chi_R| \subset \{x: R \leq |x| \leq \alpha R\}$, it follows from (4.3) that there exists a sequence $R_j \to \infty$ such that for every fixed N:

$$\lim_{R_j\to\infty} \int_{\mathbb{R}^n} u^2|\nabla\chi_{R_j}|^2 e^{2h_N} dx = 0 . \tag{4.21}$$

Letting $R \to \infty$ in (4.20) through the sequence $\{R_j\}$, taking note of (4.17) and (4.21), it follows with the aid of the dominated convergence theorem and Fatou's Lemma that

$$\int_{\mathbb{R}^n} u^2 e^{2h_N(x)} (c(x) - |\nabla h(x)|^2)\, dx \leq \int_{\mathbb{R}^n} f^2 e^{2h_N(x)} (c(x) - |\nabla h(x)|^2)^{-1} dx \tag{4.22}$$

Finally letting $N \to \infty$ in (4.22) we obtain the inequality (4.5). This completes the proof of the theorem.

In order to apply Theorem 4.1 one needs of course to find solutions of the eikonal inequality (4.4). A natural question to ask is whether there exists an optimal choice (in some sense) of the weight function h to be used in (4.5) when for instance f is a function with compact support. We shall discuss briefly this question.

Let $c(x)$ be a positive continuous function on $\mathbb{R}^n$. Introduce the Riemannian metric ds^2 on $\mathbb{R}^n$ defined by

$$ds^2 = c(x)\,|dx|^2 \tag{4.23}$$

where $|dx|^2 = dx_1^2 + \ldots + dx_n^2$. The distance between points x any y in $\mathbb{R}^n$ in this metric is given by

$$\varrho(x,y) = \inf_{\gamma} \int_0^1 c(\gamma(t))^{1/2} |\dot{\gamma}(t)|\, dt \tag{4.24}$$

where the infimum is taken over all absolutely continuous paths $\gamma : [0,1] \to \mathbb{R}^n$ such that $\gamma(0) = y$, $\gamma(1) = x$. We have the following

<u>4.2 Theorem</u>: Fix a point x^0 and define $\varrho(x) = \varrho(x,x^0)$. Then

(i) $\varrho(x)$ is a Lipschitz function on $\mathbb{R}^n$ such that $|\nabla\varrho(x)|^2 \leq c(x)$ a.e. in $\mathbb{R}^n$.

(ii) If h is a real Lipschitz function then $|\nabla h(x)|^2 \leq c(x)$ a.e., iff $h(x) - h(y) \leq \varrho(x,y)$ for all $x,y \in \mathbb{R}^n$. In particular if $h(x^0) = 0$ and $|\nabla h(x)|^2 \leq c(x)$ a.e. then $h(x) \leq \varrho(x)$ on $\mathbb{R}^n$.

For the proof of Theorem 4.2 see [1 ; Th. 1.4]. We note that in these lectures we shall use only the first part of the theorem. The proof of this part is immediate.

Using Theorem 4.2 it follows that one can take in Theorem 4.1 $h(x) = (1-\varepsilon)\varrho(x)$ for any $\varepsilon > 0$ where $\varrho(x)$ is the distance of x from the origin in the metric (4.23), $c(x)$ being the function in (4.1).

The second part of Theorem 4.2 shows that this choice of h is "almost optimal" when f has a compact support.

We shall use Theorem 4.1 to derive integral decay estimates on formal eigenfunctions of P which are defined in some neighborhood of infinity in $\mathbb{R}^n$.

<u>4.3 Theorem:</u> Let u be a solution of the equation

$$(P-\lambda)u = 0 \tag{4.25}$$

in $\Omega_R = \{x: |x| > R\}$, λ real. Suppose that

$$((P-\lambda)\varphi,\varphi) \geq \int_{\Omega_R} c(x)|\varphi|^2 dx \tag{4.26}$$

for every $\varphi \in C_o^\infty(\Omega_R)$ where $c(x)$ is some positive continuous function on $\mathbb{R}^n$. Let $\varrho(x)$ be the geodesic distance from x to the origin in the metric: $ds^2 = c(x)|dx|^2$.

If $u \in L^2(\Omega_R)$ or, more generally, if

$$\liminf_{N\to\infty}\left(N^{-2} \int_{N<|x|<\alpha N} |u|^2 dx\right) = 0 \tag{4.27}$$

for some number $\alpha > 1$, then:

$$\int_{\Omega_{R+1}} c(x) e^{2(1-\varepsilon)\varrho(x)} |u|^2 dx < \infty \tag{4.28}$$

for any $\varepsilon > 0$.

<u>Proof</u>: We first observe that in view of (4.26) and Lemma 1.7 there exists a non-negative function

$\chi \in C_o^\infty(\mathbb{R}^n)$ such that setting $P_\chi = P+\chi$, we have:

$$((P_\chi-\lambda)\varphi,\varphi) \geq \int_{\mathbb{R}^n} c(x)|\varphi|^2 dx \tag{4.29}$$

for all $\varphi \in C_o^\infty(\mathbb{R}^n)$. Next, we pick a function $\zeta \in C_o^\infty(\mathbb{R}^n)$ such that $\zeta(x) = 1$ for $|x| \leq R+\frac{1}{2}$, $\zeta(x) = 0$ for $|x| \geq R+1$ and define a function v in $\mathbb{R}^n$ as follows:

$$v(x) = (1-\zeta(x))u(x) \quad \text{for } x \in \Omega_R ,$$

$$v(x) = 0 \quad \text{for } |x| \leq R .$$

It is clear that $v \in H^1_{loc}(\mathbb{R}^n)$ and that

$$v(x) = u(x) \quad \text{for } |x| > R+1 . \tag{4.30}$$

Moreover, it follows from (4.25) and (4.30) that v is a solution of the equation:

$$(P_\chi - \lambda) v = f \tag{4.31}$$

in $\mathbb{R}^n$ where f is a function with compact support belonging to $L^2(\mathbb{R}^n)$ $(f = -2\nabla\xi \cdot \nabla u + (\chi(1-\xi) - \Delta\xi) u)$.

From (4.31), (4.29) and (4.27) it follows that v verifies the conditions of Theorem 4.1. Applying the theorem with $h(x) = (1-\varepsilon)\varrho(x)$ (using Theorem 4.2 and (4.30)) it follows that u verifies the L^2 upper bound (4.28). This completes the proof.

We conclude this section with a simple application of Theorem 4.3.

4.4 Theorem: Let H be the self-adjoint realization of P in $L^2(\mathbb{R}^n)$. Let $\psi(x)$ be an eigenfunction of H with eigenvalue $\lambda < \Sigma = \inf \sigma_{ess}(H)$. Then:

$$\int_{\mathbb{R}^n} |\psi(x)|^2 e^{2\alpha|x|} dx < \infty \tag{4.32}$$

for any $\alpha < (\Sigma - \lambda)^{1/2}$.

Proof: Let E be any number such that $\lambda < E < \Sigma$. Applying formula (1.13) it follows that there exists a neighborhood of infinity Ω_R such that

$$((P-\lambda)\varphi, \varphi) \geqslant (E-\lambda) \|\varphi\|^2$$

for all $\varphi \in C_0^\infty(\Omega_R)$. Applying to ψ Theorem 4.3 with $c(x) = E-\lambda$, $\varrho(x) = (E-\lambda)^{1/2}|x|$, we obtain (4.32) for any $\alpha < (E-\lambda)^{1/2}$. This yields the theorem.

5. Geometric spectral analysis and exponential decay of eigenfunctions

Our main goal in the following is to apply Theorem 4.3 to N-body type Schrödinger operators. To this end we shall first show how for a general Schrödinger operator P, for a given $\lambda < \Sigma(P)$, one can find a positive function $c(x)$ such that condition (4.26) holds. The function $c(x)$ will be homogeneous of degree zero and will depend on the different behavior of V in different directions in $\mathbb{R}^n$.

5.1 Definition: Let P be a Schrödinger operator on $\mathbb{R}^n$ and let $S^{n-1} = \{\omega \in \mathbb{R}^n : |\omega| = 1\}$. For $\omega \in S^{n-1}$, $0 < \varepsilon < \pi$ and $N > 0$ define

$$\Gamma_{\omega}^{\varepsilon,N} = \{x \in \mathbb{R}^n : \langle x,\omega\rangle > |x| \cos \varepsilon , |x| > N\} \tag{5.1}$$

$$\Sigma^{\varepsilon,N}(\omega) = \{\inf \frac{(P\varphi,\varphi)}{\|\varphi\|^2} : \varphi \in C_0^{\infty}(\Gamma_{\omega}^{\varepsilon,N}), \varphi \neq 0\} \tag{5.2}$$

$$K(\omega) = K(\omega;P) = \lim_{\varepsilon\to 0} \lim_{N\to\infty} \Sigma^{\varepsilon,N}(\omega) \quad . \tag{5.3}$$

In (5.1) $\langle \cdot,\cdot \rangle$ denotes the usual inner product in $\mathbb{R}^n$. $\Gamma_{\omega}^{\varepsilon,N}$ is a truncated cone with angle of opening ε. The limit in (5.3) exists because $\Sigma^{\varepsilon,N}(\omega)$ increases as $N \to \infty$ and $\varepsilon \to 0$. Note that $K(\omega)$ may take the value $+\infty$.

<u>5.2 Theorem</u>: Let $P = -\Delta + V$ satisfy the same conditions as in the preceding section. The function $K(\omega) = K(\omega;P)$ defined by (5.3) has the following properties:

(i) $K(\omega)$ is a lower semicontinuous function of ω on S^{n-1}.

(ii)
$$\min\{K(\omega) : \omega \in S^{n-1}\} = \Sigma(P) \quad . \tag{5.4}$$

(iii) Let $c(\omega)$ be a continuous function on S^{n-1} such that $c(\omega) < K(\omega)$ for all $\omega \in S^{n-1}$. Then there exists a neighborhood of infinity Ω_R in $\mathbb{R}^n$ such that

$$(P\varphi,\varphi) \geq \int_{\Omega_R} c(\frac{x}{|x|}) \, |\varphi(x)|^2 dx \tag{5.5}$$

for all $\varphi \in C_0^{\infty}(\Omega_R)$.

<u>Proof</u>: To prove (i) let $\{\omega_j\}$ be a sequence of points in S^{n-1} such that $\lim_{j\to\infty} \omega_j = \omega$. Fix a number L such that $L < K(\omega)$. From the definition of $K(\omega)$ it follows that there exist $\varepsilon \in (0, \frac{\pi}{2})$ and $N > 0$ such that $\Sigma^{\varepsilon,N}(\omega) > L$. Since $\omega_j \to \omega$ it is clear that $\Gamma_{\omega_j}^{\varepsilon/2,N} \subset \Gamma_{\omega}^{\varepsilon,N}$ for all $j \geq j_0$ for some j_0. Hence

$$K(\omega_j) \geq \Sigma^{\varepsilon/2,N}(\omega_j) \geq \Sigma^{\varepsilon,N}(\omega) > L$$

for $j \geq j_0$ which implies, since L is an arbitrary number $< K(\omega)$ that

$$\liminf_{j\to\infty} K(\omega_j) \geq K(\omega) \quad .$$

This proves that $K(\omega)$ is lower semicontinuous.

Next to prove (ii) fix $\delta > 0$ and note that from the definition of $\Sigma(P)$ it follows that there exists a ball $B = B_R(0)$ such that

$$\Sigma(P) \leq \inf\left\{\frac{(P\varphi,\varphi)}{\|\varphi\|^2} : \varphi \in C_o^\infty(\mathbb{R}^n\setminus\bar{B}), \varphi \neq 0\right\} + \delta . \tag{5.6}$$

It follows from (5.6) that

$$\Sigma(P) \leq \inf\left\{\frac{(P\varphi,\varphi)}{\|\varphi\|^2} : \varphi \in C_o^\infty(\Gamma_\omega^{\varepsilon,N}), \varphi \neq 0\right\} + \delta$$
$$= \Sigma^{\varepsilon,N}(\omega) + \delta$$

for any $\omega \in S^{n-1}$, $0 < \varepsilon < \pi/2$ and $N > R$ which implies that

$$\Sigma(P) \leq \lim_{\varepsilon\to o}\lim_{N\to\infty} \Sigma^{\varepsilon,N}(\omega) + \delta = K(\omega) + \delta$$

for any $\omega \in S^{n-1}$. Hence

$$\Sigma(P) \leq \min\{K(\omega) : \omega \in S^{n-1}\} . \tag{5.7}$$

To complete the proof of (ii) we have to show that the reverse inequality holds. This we claim follows from property (iii). Indeed, assume we have shown already that (iii) holds. Pick any number c_o such that

$$c_o < \min\{K(\omega) : \omega \in S^{n-1}\} . \tag{5.8}$$

Applying (iii) with $c(\omega) = c_o$ it follows that there exists a neighborhood of infinity $\Omega_R = \{x : |x| > R\}$ such that

$$(P\varphi,\varphi) \geq c_o\|\varphi\|^2 \tag{5.9}$$

for all $\varphi \in C_o^\infty(\Omega_R)$, which in view of (1.13) implies that

$$\Sigma(P) \geq c_o . \tag{5.10}$$

Since we can choose for c_o any number satisfying (5.8), it follows from (5.10) that

$$\Sigma(P) \geq \min\{K(\omega) : \omega \in S^{n-1}\}$$

which together with (5.7) yields (ii).

We complete the proof of the theorem by establishing (iii). Since $K(\omega) - c(\omega)$ is lower semicontinuous and positive we can choose $\delta > 0$ such that $c(\omega) + 2\delta < K(\omega)$ for all $\omega \in S^{n-1}$. Fix a point $\omega_o \in S^{n-1}$. Then there exist $0 < \varepsilon_o < \pi/2$ and $N_o > 0$ so that

$$\Sigma^{\varepsilon_o,N_o}(\omega_o) > c(\omega_o) + 2\delta . \tag{5.11}$$

It follows from (5.11) and the continuity of $c(\omega)$ that there exists a positive number $\varepsilon_1 < \varepsilon_o$ such that

$$(P\varphi,\varphi) \geq \int_{\mathbb{R}^n} (c(\frac{x}{|x|}) + \delta)\,|\varphi|^2 dx \tag{5.12}$$

for all $\varphi \in C_o^\infty(\Gamma_{\omega_o}^{\varepsilon_1,N_o})$. From the compactness of S^{n-1} it follows by a covering argument that there exist numbers $0 < \varepsilon' < \pi/2$ and $N' > 0$ such that the inequality (5.12) holds for all $\varphi \in C_o^\infty(\Gamma_\omega^{\varepsilon',N'})$ for <u>any</u> $\omega \in S^{n-1}$. We now choose a real function $\zeta(t) \in C_o^\infty(\mathbb{R})$ such that $\zeta(0) > 0$, $\zeta(t) = 0$ for $t > 1 - \cos\varepsilon'$. For any $\omega \in S^{n-1}$ and $x \in \mathbb{R}^n \setminus \{0\}$ we set

$$\zeta_\omega(x) = \zeta(1 - <\frac{x}{|x|},\omega>) . \tag{5.13}$$

Note that for every fixed ω the function ζ_ω is a C^∞ homogeneous function of degree zero on $\mathbb{R}^n \setminus \{0\}$ and that

$$\operatorname{supp} \zeta_\omega \subset \{x : <\frac{x}{|x|},\omega> \geq \cos\varepsilon'\} .$$

Let η be a positive number to be fixed later on.
Choose a number $R = R_\eta > N'$ such that

$$|\nabla\zeta_\omega(x)|^2 < \eta \text{ for } |x| > R. \tag{5.14}$$

(Note that $\nabla\zeta_\omega \to 0$ as $|x| \to \infty$.) Let φ be a given function in $C_o^\infty(\Omega_R)$. Applying the inequality (5.12) to the function $\zeta_\omega\varphi \in C_o^\infty(\Gamma_\omega^{\varepsilon',N'})$, we have:

$$\int_{\mathbb{R}^n} (|\nabla(\zeta_\omega\varphi)|^2 + V|\zeta_\omega\varphi|^2)\,dx \geq \int_{\mathbb{R}^n} (c(\frac{x}{|x|}) + \delta)\,|\zeta_\omega\varphi|^2 dx . \tag{5.15}$$

Set: $f = P\varphi$. Then $f \in L_o^1(\mathbb{R}^n)$, and integrating by parts we have:

$$\int_{\mathbb{R}^n} \nabla\varphi\cdot\nabla(\zeta_\omega^2\bar{\varphi})\,dx + \int_{\mathbb{R}^n} V\zeta_\omega^2|\varphi|^2 dx = \int_{\mathbb{R}^n} f\zeta_\omega^2\bar{\varphi} dx . \tag{5.16}$$

Using the identity

$$\operatorname{Re} \nabla\varphi\cdot\nabla(\zeta_\omega^2\bar{\varphi}) = |\nabla(\zeta_\omega\varphi)|^2 - |\varphi|^2|\nabla\zeta_\omega|^2 ,$$

it follows from (5.16) that

$$\int_{\mathbb{R}^n} (|\nabla(\zeta_\omega\varphi)|^2 + V|\zeta_\omega\varphi|^2)\,dx - \int_{\mathbb{R}^n} |\nabla\zeta_\omega|^2|\varphi|^2 dx = \operatorname{Re}\int_{\mathbb{R}^n} f\zeta_\omega^2\bar{\varphi}\,dx . \tag{5.17}$$

Combining (5.17) and (5.15), using (5.14), we get

$$\int_{\mathbb{R}^n} \left(c\left(\frac{x}{|x|}\right) + \delta\right) |\zeta_\omega \varphi|^2 dx \leq \operatorname{Re} \int_{\mathbb{R}^n} f\zeta_\omega^2 \bar{\varphi} dx + \eta \int_{\mathbb{R}^n} |\varphi|^2 dx . \tag{5.18}$$

Integrating the inequality (5.18) with respect to ω on the unit-sphere, we find that

$$\int_{\mathbb{R}^n} J(x) \left(c\left(\frac{x}{|x|}\right) + \delta\right) |\varphi|^2 dx \leq \operatorname{Re} \int_{\mathbb{R}^n} J(x) f\bar{\varphi} dx + \eta A \int_{\mathbb{R}^n} |\varphi|^2 dx \tag{5.19}$$

where

$$J(x) = \int_{S^{n-1}} \zeta_\omega(x)^2 d\omega \, , \quad A = \int_{S^{n-1}} d\omega \, , \tag{5.20}$$

$d\omega$ denoting the induced Lebesgue measure on $S^{n-1} \subset \mathbb{R}^n$.

It follows from (5.20) and (5.13) that $J(x)$ is a positive homogeneous function of degree zero on $\mathbb{R}^n \setminus \{0\}$. It is also clear that $J(x)$ is invariant under rotations in $\mathbb{R}^n$. These properties imply that $J(x) = J_o$ where J_o is a positive constant. We now fix the constant η in (5.14), choosing: $\eta = \delta J_o / A$. Using the last remark and recalling the definition of f, it thus follows from (5.19) that

$$\int_{\mathbb{R}^n} c\left(\frac{x}{|x|}\right) |\varphi|^2 dx \leq \operatorname{Re} \int_{\mathbb{R}^n} f\bar{\varphi}\, dx = (P\varphi, \varphi)$$

for every $\varphi \in C_o^\infty(\Omega_R)$. This yields (iii) and completes the proof of the theorem.

In the following theorems we will have to consider distance functions ϱ_c defined by a Riemannian metric

$$ds_c^2 = c(x)\,|dx|^2$$

where $c(x)$ is a lower semicontinuous but not necessarily continuous function. As before we define:

$$\varrho_c(x,y) = \inf_\gamma \int_o^1 c(\gamma(t))^{1/2} |\dot{\gamma}(t)|\, dt$$

where the infimum is taken over all absolutely continuous paths $\gamma : [0,1] \to \mathbb{R}^n$ which join y to x. We shall need the following convergence result for a sequence of metrics ϱ_{c_j} with $c_j \uparrow c$.

<u>5.3 Lemma</u>: Let $\{c_j(x)\}$, $j = 1,2,\ldots$, be a non-decreasing sequence of non-negative, locally bounded lower semicontinuous functions on $\mathbb{R}^n$, such that $c_j(x) \geq \delta > 0$, except for possibly finitely many x. Set $c(x) = \lim_{j\to\infty} c_j(x)$ and suppose that $c(x)$ is locally bounded. Then

$c(x)$ is lower semicontinuous and

$$\lim_{j\to\infty} \varrho_{c_j}(x,y) = \varrho_c(x,y)$$

uniformly in (x,y) in any compact subset of $\mathbb{R}^n \times \mathbb{R}^n$.

For the proof of Lemma 5.3 we refer to [1 ; Lemma 4.3].

<u>5.4 Theorem</u>: Let $P = -\Delta + V$ satisfy the hypotheses of Theorem 1.1. Let u be a solution of the differential equation $(P-\lambda)u = 0$ in some neighborhood of infinity Ω in $\mathbb{R}^n$. Suppose that $\lambda < \Sigma(P)$ and that the function $K(\omega) = K(\omega;P)$ (defined by (5.3)) is bounded on S^{n-1}. Set: $c(x) = K(x/|x|) - \lambda$ if $x \neq 0$, $c(0) = 0$. If $u \in L^2(\Omega)$, then:

$$\int_\Omega |u(x)|^2 e^{2(1-\varepsilon)\varrho_c(x)} dx < \infty \tag{5.21}$$

for any $\varepsilon > 0$ where $\varrho_c(x)$ denotes the geodesic distance from x to the origin in the metric: $ds^2 = c(x)|dx|^2$.

<u>Proof</u>: The first step in the proof is to approximate ϱ_c by some ϱ_{c_j} where c_j is continuous on $\mathbb{R}^n \setminus \{0\}$. Let $\{c_j(\omega)\}$ be a non-decreasing sequence of continuous functions on $S^{n-1} \subset \mathbb{R}^n$ such that $0 < c_j(\omega) < c(\omega)$ and $\lim_{j\to\infty} c_j(\omega) = c(\omega)$ for all $\omega \in S^{n-1}$. (A sequence with these properties exists by the lower semicontinuity of $c(\omega)$.) We also denote by c_j the extensions of $c_j(\omega)$ to $\mathbb{R}^n$ defined by $c_j(x) = c_j(x/|x|)$ if $x \neq 0$, $c_j(0) = 0$. The sequence $c_j(x)$ satisfies the hypotheses of Lemma 5.3. Therefore

$$\lim_{j\to\infty} \varrho_{c_j}(x) = \varrho_c(x)$$

uniformly on compact subsets on $\mathbb{R}^n$ where $\varrho_{c_j}(x)$ denotes the geodesic distance from x to the origin in the metric: $ds^2 = c_j(x)|dx|^2$. Therefore, given $0 < \varepsilon < 1$ there exists j_o such that

$$(1-\tfrac{\varepsilon}{2})\varrho_c(\omega) \leqslant \varrho_{c_{j_o}}(\omega) \leqslant \varrho_c(\omega)$$

for $\omega \in S^{n-1} \subset \mathbb{R}^n$. It follows from their definition that the $\varrho_{c_j}(x)$ are homogeneous functions on $\mathbb{R}^n$. Therefore

$$(1-\tfrac{\varepsilon}{2})\varrho_c(x) \leqslant \varrho_{c_{j_o}}(x) \leqslant \varrho_c(x) \tag{5.22}$$

for every $x \in \mathbb{R}^n$.

Next observe that since $c_{j_o}(\omega) < c(\omega) = K(\omega;P)-\lambda = K(\omega;P-\lambda)$, it follows from Theorem 5.2 that there exists a neighborhood of infinity Ω_R in $\mathbb{R}^n$ such that

$$((P-\lambda)\varphi,\varphi) \geq \int_{\Omega_R} c_{j_o}(x)\,|\varphi(x)|^2 dx \tag{5.23}$$

for every $\varphi \in C_o^\infty(\Omega_R)$.

We are now in a position to apply Theorem 4.3 to the given function u which is an L^2 solution of the equation $(P-\lambda)u = 0$ in some neighborhood of infinity Ω . Assuming as we may that $\Omega \supset \bar{\Omega}_R$ and noting that on Ω_R the function $c_{j_o}(x)$ is bounded from below by some positive constant δ , it follows from Theorem 4.3 that

$$\int_{\Omega_R} |u(x)|^2 \exp((2-\varepsilon)\varrho_{c_{j_o}}(x))\,dx < \infty , \tag{5.24}$$

$\varepsilon > 0$ the number which was fixed above. Combining (5.24) and (5.22), noting that $u \in L^2(\Omega)$, we obtain the L^2 upperbound (5.21). This completes the proof.

6. N-body Schrödinger operators

We shall need the following

6.1 Lemma: Let $P = -\Delta + V$ satisfy the hypotheses of Theorem 1.1. Let $\Lambda(P)$, $\Sigma(P)$ and $K(\omega) = K(\omega;P)$ be as defined in (1.12), (1.13) and (5.3). Suppose that $V(x+t\omega_o) = V(x)$ for some $\omega_o \in S^{n-1} \subset \mathbb{R}^n$ and all $t \in \mathbb{R}$. Then:

(i) $\Lambda(P) = \Sigma(P)$.

(ii) $K(\omega_o{:}P) = \Sigma(P) = \min\{K(\omega) : \omega \in S^{n-1}\}$.

Proof: Let $\varphi \in C_o^\infty(\mathbb{R}^n)$ and define $\varphi_\tau(x) = \varphi(x+\tau\omega_o)$ for $\tau \in \mathbb{R}$. Then

$$\frac{(P\varphi,\varphi)}{\|\varphi\|^2} = \frac{(P\varphi_\tau,\varphi_\tau)}{\|\varphi_\tau\|^2} .$$

To prove (i) we need only to show that $\Sigma(P) \leq \Lambda(P)$, since the opposite inequality follows from the definitions. Fix $\varepsilon > 0$ and let K be a compact set such that

$$\inf\left\{\frac{(P\varphi,\varphi)}{\|\varphi\|^2} : \varphi \in C_o^\infty(\mathbb{R}^n\setminus K)\right\} > \Sigma - \varepsilon .$$

Pick $\varphi \in C_o^\infty(\mathbb{R}^n)$ such that

$$\frac{(P\varphi,\varphi)}{\|\varphi\|^2} < \Lambda(P)+\varepsilon \ .$$

Since φ has compact support there exist τ such that $\varphi_\tau \in C_0^\infty(\mathbb{R}^n\setminus K)$. Thus

$$\Lambda(P) + \varepsilon > \frac{(P\varphi,\varphi)}{\|\varphi\|^2} = \frac{(P\varphi_\tau,\varphi_\tau)}{\|\varphi_\tau\|^2}$$

$$\geqslant \inf\left\{\frac{(P\varphi,\varphi)}{\|\varphi\|^2} : \varphi \in C_0^\infty(\mathbb{R}^n\setminus K)\right\} > \Sigma - \varepsilon$$

which proves (i) since $\varepsilon > 0$ was arbitrary.

Noting that for any $\varphi \in C_0^\infty(\mathbb{R}^n)$, φ_τ is supported in the truncated cone $\Gamma_{\omega_0}^{\varepsilon,N}$ (see Definition (5.1)) for any ε and N when τ is large enough, the first equality in (ii) follows from a similar argument. The second equality is equation (5.4) so the proof is complete.

We now consider a class of Schrödinger operators which contains various Schrödinger operators which arise in the study of N-body quantum systems.

<u>6.2 Definition</u>: Let $P = -\Delta + V$ be a Schrödinger operator on $\mathbb{R}^n$. We shall say that P is a multiparticle type Schrödinger operator if

$$V(x) = \sum_{i=1}^{\ell} V_i(x) \tag{6.1}$$

where $V_i(x)$ are real functions on $\mathbb{R}^n$ having the following properties. There exist non-zero projections $\Pi_i : \mathbb{R}^n \to \mathbb{R}^n$, $i = 1,\dots,\ell$, $\Pi_i \neq \Pi_j$ for $i \neq j$, such that

(i) $V_i(x) = V_i(\Pi_i x)$.

(ii) $V_i(x) \to 0$ as $|\Pi_i x| \to \infty$.

Also, set $Y_i = \operatorname{Ran} \Pi_i$, $m_i = \dim Y_i > 0$. Then

(iii) $V_i \restriction Y_i \in L^1_{loc}(Y_i)$, $(V_i)_- \restriction Y_i \in L^{p_i}_{loc}(Y_i)$ with $p_i > m_i/2$, $i = 1,\dots,\ell$.

Using Lemma 0.2 it is easy to see that if V verifies the conditions of Definition 6.2 then for any $\varepsilon > 0$ the following inequality holds:

$$\|V_-^{1/2}\varphi\| \leqslant \varepsilon\|\nabla\varphi\| + C_\varepsilon\|\varphi\|$$

for any $\varphi \in C_o^\infty(\mathbb{R}^n)$ where C_ε is some constant ($\|\cdot\|$ denotes the norm in $L^2(\mathbb{R}^n)$). It thus follows that $P = -\Delta + V$ verifies the conditions of Theorem 1.1.

Next we compute the function $K(\omega;P)$ (defined by (5.3)).

<u>6.3 Theorem</u>: Let $P = -\Delta + \sum_{i=1}^{\ell} V_i$ be a multiparticle type Schrödinger operator. Then for any $\omega \in S^{n-1} \subset \mathbb{R}^n$:

$$K(\omega;P) = K(\omega;P_\omega) = \Sigma(P_\omega) = \Lambda(P_\omega) \tag{6.2}$$

where

$$P_\omega = -\Delta + \sum_{\Pi_i\omega=0} V_i . \tag{6.3}$$

(The last sum denotes summation of V_i over those indices i for which $\Pi_i\omega = 0$, the sum being defined as 0 when $\Pi_i\omega \neq 0$ for all i.)

<u>Proof</u>: Consider a truncated cone $\Gamma_\omega^{\varepsilon,N}$ in the direction of ω (see (5.1)). Suppose that $\Pi_i\omega \neq 0$. It follows from our assumptions on V_i that if $\varepsilon > 0$ is sufficiently small then $V_i(x) \to 0$ as $x \to \infty$ in $\Gamma_\omega^{\varepsilon,N}$. It is thus clear from the definition of $K(\omega;P)$ that V_i can be deleted from the sum $V = \sum_{i=1}^{\ell} V_i$ when calculating $K(\omega;P)$. This gives that

$$K(\omega;P) = K(\omega;P_\omega) . \tag{6.4}$$

Next observe that if $\Pi_i\omega = 0$, then

$$V_i(x+t\omega) = V_i(\Pi_i(x+t\omega)) = V_i(\Pi_i x) = V_i(x) .$$

Thus we are in a position to apply Lemma 6.1 to P_ω which gives

$$K(\omega;P_\omega) = \Sigma(P_\omega) = \Lambda(P_\omega)$$

which together with (6.4) proves the theorem.

It follows from Theorem 6.3 that the function $K(\omega;P)$ takes only a finite number of values on S^{n-1}. We also have that

$$\Sigma(P) = \min_{S^{n-1}} K(\omega;P) \leq \max_{S^{n-1}} K(\omega;P) = 0 . \tag{6.5}$$

Indeed, the only result which needs proving is that $\max K(\omega;P) = 0$. This however follows from the lower semicontinuity of $K(\omega;P)$ and the fact that $K(\omega;P) = 0$ on a dense set of points in S^{n-1}, namely the set $E = \{\omega \in S^{n-1} : \Pi_i\omega \neq 0 \text{ for } i = 1,\ldots,\ell\}$, since by Theorem 6.3:

$$K(\omega;P) = \Lambda(-\Delta) = 0 \quad \text{for} \quad \omega \in E .$$

Combining Theorem 5.4 and Theorem 6.3, using (1.15) and (6.5) we obtain the following theorem on exponential decay of eigenfunctions of multiparticle type Schrödinger operators.

6.4 Theorem: Let $P = -\Delta + \sum_{i=1}^{\ell} V_i$ be a multiparticle type Schrödinger operator on $\mathbb{R}^n$ satisfying the conditions of Definition 6.2. Let H be the self-adjoint realization of P in $L^2(\mathbb{R}^n)$ and let H_ω be the self-adjoint realization of $P_\omega = -\Delta + \sum_{\Pi_i \omega = 0} V_i$, $\omega \in S^{n-1} \subset \mathbb{R}^n$. Set

$$\Sigma_\omega = \inf \sigma(H_\omega) .$$

Let ψ be an eigenfunction of H with eigenvalue $\lambda < \inf \sigma_{ess}(H)$. Then

$$\int_{\mathbb{R}^n} |\psi(x)|^2 e^{2(1-\varepsilon)\varrho(x)} dx < \infty \tag{6.6}$$

for any $\varepsilon > 0$ where $\varrho(x)$ is the geodesic distance from x to 0 in the Riemannian metric:

$$ds^2 = (\Sigma_{x/|x|} - \lambda)|dx|^2 . \tag{6.7}$$

The L^2 exponential decay estimate of Theorem 6.4 can be converted into a <u>pointwise</u> estimate. We have the following

6.5 Theorem: Under the same hypotheses as in Theorem 6.4 and with the same notation, there exists for any $\varepsilon > 0$ a constant C_ε such that

$$|\psi(x) \leq C_\varepsilon e^{-(1-\varepsilon)\varrho(x)} \quad \text{a.e. on} \quad \mathbb{R}^n . \tag{6.8}$$

For the derivation of (6.8) from (6.6) we refer to [1 ; Theorem 5.1].

As an example we shall apply the above results to the Schrödinger operator of an atom consisting of a nucleus and N electrons with coordinates $x^i \in \mathbb{R}^3$, $i = 1,\dots,N$. The Schrödinger operator P_N of the system acts on functions defined on the configuration space $\mathbb{R}^{3N}$ with generic point $x = (x^1,\dots,x^N)$. It has the form:

$$\begin{aligned} P_N &= -\sum_{i=1}^{N} \Delta_i + \sum_{i=1}^{N} v_0(x^i) + \sum_{1\leq i<j\leq N} v(x^i - x^j) \\ &= -\Delta + V(x) \end{aligned} \tag{6.9}$$

where Δ_i is the Laplacian in x^i and where the interacting potentials

$v_o(y)$ and $v(y)$ are real functions belonging to $L^1_{loc}(\mathbb{R}^3)$. We shall assume that the following conditions hold:

(i) $|v_o(y)| + |v(y)| \to 0$ as $|y| \to \infty$.

(ii) $v(y) \geq 0$ on $\mathbb{R}^3$.

(iii) $(v_o(y))_- \in L^p_{loc}(\mathbb{R}^3)$ for some $p > 3/2$.

Under the above conditions it is readily seen that P_N belongs to the class of multiparticle Schrödinger operators introduced in Definition 6.2. In the following we assume that v and v_o are fixed and allow N to vary, we set: $P_1 = -\Delta_1 + v_o$.

We denote by H_N the self-realization of P_N in $L^2(\mathbb{R}^{3N})$. We set:

$$\Lambda_N = \inf \sigma(H_N) \tag{6.10}$$

for $N = 1,2,\ldots,\Lambda_o = 0$. For any fixed N and $x = (x^1,\ldots,x^N) \in \mathbb{R}^{3N}$, we define:

$$I_N(x) = \text{number of } x^i,\ 1 \leq i \leq N,\ \text{such that } x^i = 0. \tag{6.11}$$

Applying Theorem 6.3 to P_N, using the lower semicontinuity of $K(\omega;P_N)$ and the assumption that $v \geq 0$, it follows readily that

$$K(\omega;P_N) = \Lambda_{I_N(\omega)} \tag{6.12}$$

for $\omega \in S^{3N-1} \subset \mathbb{R}^{3N}$.

It follows from (6.12) and (6.11) that the function $K(\omega;P_N)$ takes only the finite number of values Λ_i for $i = 0,1,\ldots,N-1$. Invoking formulas (5.4) and (1.15), it follows that

$$\inf \sigma_{ess}(H_N) = \min_{\omega} K(\omega;P_N) = \min_{0 \leq i \leq N-1} \Lambda_i . \tag{6.13}$$

Using induction on N it follows from (6.13) and (6.10) that the sequence $\{\Lambda_N\}$ is non-increasing. We thus obtain

<u>6.6 Theorem</u>: Let H_N be the self-adjoint realization of the N-electron atom Schrödinger operator introduced above. The following holds:

(i) $\inf \sigma(H_N) \leq \inf \sigma(H_{N-1})$,

(ii) $\inf \sigma_{ess}(H_N) = \inf \sigma(H_{N-1})$

for $N = 1,2,\ldots$.

Theorem 6.6 gives the main part of the HVZ Theorem (see [6 ; Theorem XIII.17]).

We now apply Theorem 6.5 to the Schrödinger operator P_N. Taking note of (6.12) and Theorem 5.4 we obtain

6.7 Theorem: Let ψ be an eigenfunction of H_N with eigenvalue $\lambda < \inf \sigma_{ess}(H_N)$. Then for any $\varepsilon > 0$ there exists a constant C_ε such that

$$|\psi(x)| \leq C_\varepsilon e^{-(1-\varepsilon)\varrho(x)} \tag{6.14}$$

a.e. in $\mathbb{R}^{3N}$ where $\varrho(x)$ is the geodesic distance from $x = (x^1,\ldots,x^N)$ to the origin in $\mathbb{R}^{3N}$ in the Riemannian metric:

$$ds^2 = (\Lambda_{I_N}(x)-\lambda)|dx|^2 .$$

We conclude these lectures by mentioning some results on lower bounds which one can derive for eigenfunctions which are positive in a neighborhood of infinity. In the case of the 2-body problems such results were derived in section 3. In the case of general multiparticle type Schrödinger operators we have the following

6.8 Theorem: Let $P = -\Delta + \sum_{i=1}^{\ell} V_i$ be a multiparticle type Schrödinger operator where the V_i satisfy the conditions of Definition 6.2 with condition (iii) replaced by the stronger condition:

(iii)' $V_i \upharpoonright Y_i \in L^{p_i}_{loc}(Y_i)$ where $p_i > m_i/2$, $i = 1,\ldots,\ell$.

Let H be the self-adjoint realization of P in $L^2(\mathbb{R}^n)$ and let ψ be an eigenfunction of H with eigenvalue $\lambda < \inf \sigma_{ess}(H)$. Suppose that $\psi(x) > 0$ for $|x| \geq R$. Then the following lower bound holds:

$$\psi(x) \geq c_\varepsilon e^{-(1-\varepsilon)\varrho(x)} \quad \text{for } |x| \geq R , \tag{6.15}$$

for any given $\varepsilon > 0$ and some constant $c_\varepsilon > 0$, where $\varrho(x)$ is the geodesic distance from x to the origin in the Riemannian metric (6.7) as defined in Theorem 6.4.

For the ground state of a proper N-body Schrödinger operator the lower bound (6.15) was established by Carmona and Simon [3] using a probabilistic method. We mention that Theorem 6.8 can be proved by standard P.D.E. techniques using some comparison and positivity results of the type discussed in section 2. It seems likely that the theorem can also be proved by extending the techniques of [3].

Theorem 6.8 shows that under somewhat stronger assumptions on the V_i the upper bounds given in Theorem 6.5 (and Theorem 6.4) are "almost optimal" for eigenfunctions which are positive in a neighborhood of infinity. In particular when ψ is the ground state of a multiparticle Schrödinger operator satisfying the conditions of Theorem 6.8 it follows from Theorem 6.5 and Theorem 6.8 that one has the asymptotic relation:

$$\lim_{|x|\to\infty} \frac{\log \psi(x)}{\varrho(x)} = -1$$

where $\varrho(x)$ is the distance from x to the origin in the appropriate Riemannian metric.

References

[1] S. Agmon, Lectures on Exponential Decay of Solutions of Second-Order Elliptic Equations, Princeton University Press, 1982.

[2] S. Agmon, Lectures on Elliptic Boundary Value Problems, Van Nostrand, Princeton, 1965.

[3] R. Carmona and B. Simon, Pointwise bounds on eigenfunctions and wave packets in N-body quantum systems, V : Lower bounds and path integrals, Comm. Math. Phys. 80(1981), 59-98.

[4] D. Gilbarg and N.S. Trudinger, Elliptic Partial Differential Equations of Second Order, Springer-Verlag, Berlin and New York, 1977.

[5] A. Persson, Bounds for the discrete part of the spectrum of a semi-bounded Schrödinger operator, Math. Scand. 8(1960), 143-153.

[6] M. Reed and B. Simon, Methods of Modern Mathematical Physics, IV, Analysis of Operators, Academic Press, New York, 1978.

[7] G. Stampacchia, Le problème de Dirichlet pour les équations elliptiques du second ordre à coefficients discontinus, Ann. Inst. Fourier 15 (1965), 189-258.

The Hebrew University, Jerusalem

QUANTUM SCATTERING THEORY FOR TWO- AND THREE-BODY SYSTEMS WITH POTENTIALS OF SHORT AND LONG RANGE

Volker Enss *

Oct.84-March 85:
Division of Physics, Mathematics
and Astronomy
California Institute of Technology
Pasadena, CA 91125, USA

permanent address:
Institut für Mathematik I
Freie Universität
Arnimallee 2-6
1000 Berlin 33, West Germany

Abstract

We give a full proof of asymptotic completeness for Schrödinger operators of two- and three-particle quantum systems. The interaction is given by pair potentials which may be of short and of long range, including Coulomb forces. We apply geometrical time-dependent methods where propagation of scattering states in phase space and in configuration space is essential. The main new results are the inclusion of long-range potentials of the two-body estimates in Section VI and for three-particle systems. But also where we recover known results some of our methods are new. Where possible we have chosen the methods which admit generalization to higher particle numbers.

---

*Sherman Fairchild Scholar

Table of Contents

I. Introduction.

We present in detail the geometrical, time-dependent treatment of scattering theory for two- and three-body quantum systems. The main goal is to prove asymptotic completeness. The potentials may be of long range. One of the major tools is the study of propagation properties of scattering states in configuration space and in phase space. A state in the point spectral subspace of the Hamiltonian H which generates the internal time evolution, is a bound state. Let $\|x\|$ be any measure for the size of the system, e.g. the sum of the distances between the particles. (We have separated off the trivial center of mass motion.) For any admissible error $\varepsilon > 0$ and a given state Ψ there is a cutoff radius $R = R(\varepsilon, \Psi)$ such that the state is localized up to an error ε inside $|x| \leq R$. If Ψ is an eigenstate, $H\Psi = E\Psi$, $E \in \mathbb{R}$, then its time evolution is a trivial phase factor $\exp(-iHt)\Psi = \exp(-iEt)\Psi$ and the localization remains valid uniformly in time. Similarly the localization is uniform in time for finite or infinite superpositions of eigenvectors, i.e. any state in the point spectral subspace $\mathcal{H}^{pp}(H)$ of the Hamiltonian is geometrically a bound state, a physical bound state.

Scattering theory deals with the scattering states which are orthogonal to all bound states. In mathematical terms they belong to the continuous spectral subspace $\mathcal{H}^{cont}(H)$ of the Hamiltonian. Experience from scattering experiments suggests that any scattering state asymptotically in time decays into subsystems which move independently of each other. For two body systems this means that the interaction between the particles is weak at late and very early times, i.e. the free time evolution dominates. If the forces between the particles decay sufficiently fast with the separation of the particles (short-range potentials), then this can be expressed mathematically as

$$\lim_{\tau\to\infty} \sup_{t\geq 0} \|(e^{-iHt} - e^{-iH_0 t})\, e^{-iH\tau}\Psi\| = 0. \tag{1.1}$$

At a sufficiently late time τ (when the scattering is over) the future time evolution of the state is independent of the potential, the interacting time evolution exp(-iHt), $t \geq 0$, is indistinguishable from the free time evolution $\exp(-iH_0 t)$ on an "old" state $\exp(-iH\tau)\Psi$. In the corresponding statement for the past the supremum has to be taken for negative times $t \leq 0$ and the limit is $\tau \to -\infty$. If long-range potentials are present like the physically important Coulomb potential between charged particles, then the free time evolution has to be replaced in (1.1) by a better approximation $U(t+\tau,\tau)$, a "modified free" time evolution. This is an effect from classical mechanics, it is well known for a long time in the Kepler problem: although a hyperbolic orbit is well approximated by a straight line, the longitudinal motion differs logarithmically from one with constant speed. The modification U takes care of this effect for semi-infinite time intervals (for a definition see Section V) and one has

$$\lim_{\tau\to\infty} \sup_{t\geq 0} \|[e^{-iHt} - U(t+\tau,\tau)]\, e^{-iH\tau}\Psi\| = 0. \tag{1.2}$$

For finite time intervals (for experiments carried out in a finite time) the modification can be omitted.

$$\lim_{\tau\to\infty} \sup_{|t|\leq T} \|[e^{-iHt} - e^{-iH_0 t}]\, e^{-iH\tau}\Psi\| = 0 \tag{1.3}$$

for any $T < \infty$, even if long range forces are present.

The physical intuitive results (1.1) and (1.2) which are called "asymptotic completeness" imply that the continuous part of the Hamiltonian $H\restriction\mathcal{H}^{cont}(H)$ is unitarily equivalent to H_0. Since the latter (Laplacian on Euclidean space $\mathbb{R}^\nu$) is completely known we have good control of the spectrum etc. The corresponding statements for three-body systems will be given in Section VII when some notation has been introduced.

The particles are moving in Euclidean space $\mathbb{R}^\nu$, the dimension ν is arbitrary. We separate off the trivial free motion, then $x \in \mathbb{R}^\nu$

denotes the relative position of the particles, p their relative momentum and $m = m_1 \cdot m_2/(m_1+m_2)$ their reduced mass. $\mathcal{H} = L^2(\mathbb{R}^\nu)$ is the state space. For simplicity we discuss here the standard case where the free Hamiltonian H_0 is

$$(1.4) \qquad H_0 = \frac{p^2}{2m} = -\frac{1}{2m}\Delta .$$

It is self-adjoint on the domain $\mathfrak{D}(H_0) \subset \mathcal{H}$. With minor changes the method and results carry over to a very wide class of functions $h_0(p)$. The interacting Hamiltonian H is the operator sum

$$(1.5) \qquad H = H_0 + V .$$

With some extra work on domain questions everything can be easily extended to form sums of more singular operators. Here we assume that V be a Kato-bounded multiplication operator: $\mathfrak{D}(V) \supset \mathfrak{D}(H_0)$, and there are $a < 1$, $b < \infty$ with

$$(1.6) \qquad \|V\Psi\| \le a\|H_0\Psi\| + b\|\Psi\| \quad \forall\Psi \in \mathfrak{D}(H_0).$$

Then H is self-adjoint by the Kato-Rellich theorem on $\mathfrak{D}(H_0)$. (Non-local velocity dependent potentials could be included easily.) The potential may consist of a short-range part V_s and a bounded part V_ℓ of long range. We denote by

$$F(|x|>R)$$

the multiplication operator in x-space with the characteristic function of the indicated region and similarly for spectral projections of other self-adjoint operators like $F(E_1 < H < E_2)$. Then we can express our decay requirements as follows.

$$(1.7) \qquad \|(H_0+\mathbb{1})^{-1} V_s F(|x|>R)\| \in L^1(\mathbb{R}_+, dR).$$

This is equivalent to

$$(1.7') \qquad \|V_s (H_0+\mathbb{1})^{-1} F(|x|>R)\| \in L^1(\mathbb{R}_+, dR).$$

The long-range potential is given by the differentiable function $V_\ell(x)$ which tends to zero as $|x| \to \infty$. Its derivative should satisfy

$$(1.8) \qquad |(\nabla V_\ell)(x)| \leq \text{const } (1+|x|)^{-(\delta+3/2)}, \quad \delta > 0.$$

A trivial consequence is

$$|V_\ell(x)| \leq \text{const}(1+|x|)^{-(\delta+1/2)}.$$

The Coulomb potential with $\delta = 1/2$ is admitted. Clearly the splitting of a given potential is not unique. If a long-range part $\tilde{V}_\ell$ satisfies (1.8) then by Lemma 3.3 of Hörmander [22] it can be decomposed into a smooth part and a remainder of short range. The smooth long-range part V_ℓ satisfies for all multiindices α with $|\alpha| \geqslant 1$

$$(1.9) \qquad |(D^\alpha V_\ell(x)| \leq \text{const}(\alpha)\cdot(1+|x|)^{-1-\delta-|\alpha|/2},$$

and the remainder satisfies (1.7). Thus one can require (1.9) without loss of generality. It is *not* an additional condition on the potential V, it states only that a suitable splitting is chosen among all possible ones.

Our results, in particular asymptotic completeness are proved under these assumptions. One gets a better decay rate in Theorem 6.1 if condition (1.7) is replaced by the slightly stronger

$$(1.10) \qquad \|(H_0+\mathbb{1})^{-1} V_s F(|x|\geqslant R)\| \leq \text{const}(1+R)^{-1-\varepsilon}, \quad \varepsilon > 0.$$

Both conditions (1.7) and (1.10), respectively, can be weakened by replacing $(H_0+\mathbb{1})^{-1} V_s$ with $g(H_0)V_s$ and requiring the given decay for any $g \in C_0^\infty(\mathbb{R})$. The proof can remain essentially unchanged.

For later use we collect a few elementary consequences of our assumptions. Relative boundedness of H and H_0

$$(1.11) \qquad \|H_0 (H-z')^{-1}\| + \|H (H_0-z)^{-1}\| < \infty$$

for z, z' in the resolvent sets and

$$\|V (H-z')^{-1}\| + \|V (H_0-z)^{-1}\| < \infty \tag{1.12}$$

follow from (1.6). Let $h(x)$ denote any bounded function with $\lim h(x) = 0$ as $|x| \to \infty$. Then by elementary Fourier analysis

$$h(x)\ (H_0-z)^{-1/2} \tag{1.13}$$

is a compact operator. The difference of the resolvents

$$\begin{aligned} &(H_0-z)^{-1} - (H-z)^{-1} \\ &= (H_0-z)^{-1}\ V\ (H-z)^{-1} \\ &= (H_0-z)^{-1}(1+|x|)^{-1/2} \cdot (1+|x|)^{1/2}\ V\ (H-z)^{-1} \end{aligned} \tag{1.14}$$

is compact as a product of a compact operator with one that is bounded by (1.7), (1.8). Similarly

$$\begin{aligned} &(H_0-z)^{-1/2}(1+|x|)\ V_s\ (H-z)^{-1} \\ &= (H_0-z)^{-1/2}\ F(|x|<R)\ (1+|x|)\ V_s\ (H-z)^{-1} \\ &+ (H_0-z)^{-1/2}\ F(|x|>R)\ (1+|x|)\ V_s\ (H-z)^{-1} \end{aligned} \tag{1.15}$$

is compact as a sum of a compact operator for any R and an operator with arbitrary small norm. It follows from the integrability (1.7) and the monotonicity that

$$(1+R)\ \|F(|x|>R)\ V_s\ (H-z)^{-1}\| \to 0 \text{ as } R \to \infty. \tag{1.16}$$

Our mathematical treatment of scattering theory follows closely the guidelines of physical intuition. When the scattering is over one expects that the particles move away from each other and that their separation becomes parallel to their relative velocity. Certain regions of phase space are "absorbing" for particle trajectories even under the influence of forces. We prove that in Section II. It is remarkable that one can get these results without any detailed analysis of the interacting time evolution. Only kinematical relations between suitably chosen observables play a rôle. The effects of the forces are nothing more than a small perturbation which disappears for large times. This can be true although the evolution of the state depends strongly on the potentials because we have asked only "modest questions" about the phase space localization.

The "modest answers" are sufficient to show that the future true time evolution is well approximated by simpler ones on the absorbing subsets of the state space. In Section III we treat short-range potentials and use the free time evolution as an approximation. If long-range forces are present they have an effect on the motion of the particle even asymptotically. In Section IV we introduce an "intermediate" approximate time evolution and discuss its properties. It is sufficiently close to the true evolution such that it is easy to control the error. On the other hand it is sufficiently simple to yield simple answers to "less modest" questions about the localization in phase space. The trick is to decompose $\exp[-i(H_0+V(x))t] \approx \exp[-iH_0t] \times \exp[-iV(x)t]$ for longer and longer time intervals on suitable states. Thus either p or x change with time but not both simultaneously. This simplifies the estimates very much. The better control of the localization in phase space is used in Section V to show that Dollard's modified free time evolution is a good approximation of the true one at large times. This finishes the proof of asymptotic completeness for short- and long-range potentials. If the only long-range potentials are Coulomb forces then the introduction of an "intermediate" time evolution can be omitted.

Section VI goes beyond asymptotic completeness for two-body systems. A state which is localized inside a ball of radius R and has energy below $mv^2/2$ should be at time t inside a ball of radius R + v|t|. We show that the "tails" of the wavefunction outside that ball decay integrably or rapidly in t. The result will be used in the treatment of three-body systems, it is important there that v can be chosen small.

The three-body problem will be treated in Part B of these notes. The introduction for that is contained in Section VII, general references are given in the last section.

Part A. *Two-Body Systems*

II. Asymptotic Observables and Propagation in Phase Space.

We study the propagation of scattering states in phase space under the interacting time evolution. The following two theorems are the main results of this section.

Theorem 2.1. Let $H = H_0 + V_s + V_\ell$ satisfy (1.4)-(1.8) and let $\Psi \in \mathcal{H}^{cont}(H)$, $f \in C_0^\infty(\mathbb{R}^\nu)$, then

$$\lim_{\tau\to\infty} \|[f(\tfrac{x}{\tau}) - f(\tfrac{p}{m})]\, e^{-iH\tau}\, \Psi\| = 0. \tag{2.1}$$

It says that for any scattering state the distribution of values for the family of self-adjoint operators x/τ (average velocity) asymptotically coincides with the distribution of velocities p/m at time τ. For a free particle it is easy to see that

$$\frac{x}{\tau} - \frac{p}{m} \sim \tau^{-1}; \tag{2.2}$$

we restrict ourselves here to the far weaker statement that the difference goes to zero. The advantage is that this modest question can be answered easily for an extremely wide class of interactions, no detailed information is used about the interacting time evolution.

To apply this result to scattering theory we introduce the following decomposition of the identity on a spherical shell. For any given pair of lower and upper energy cutoffs $0 < E_1 < E_2 < \infty$ there is a finite family $\{f_i\}$, $f_i \in C_0^\infty(\mathbb{R}^\nu)$, $0 \le f_i(w) \le 1$, such that

$$\text{supp } f_i \subset \{w \in \mathbb{R}^\nu \,\big|\, |w-v_i| < |v_i|/2\} \tag{2.3}$$

for a suitable finite collection of velocities $0 \ne v_i \in \mathbb{R}^\nu$, and

$$\sum_i f_i^2(w) = 1 \text{ if } E_1 \le mw^2/2 \le E_2. \tag{2.4}$$

The support condition (2.3) is chosen such that for any i all directions in the support of f_i enclose an acute angle. A state in the range of the operator $f_i(x/\tau)$ is localized outside a ball of radius $|v_i| \cdot \tau/2$. For large τ the localization is approximately the same for states in the range of $f_i(p/m)\ f_i(x/\tau)$. Now in addition the velocities point roughly in the same direction as the position does. The states are "far from the origin and outgoing". It is a simple consequence of Theorem 2.1 that in the far future any scattering state is well approximated by a finite collection of pieces which are far from the scatterer and outgoing. On these components the future time evolution is simple as we will see in the next sections.

Theorem 2.2. Let $\Psi \in \mathcal{H}^{\text{cont}}(H)$ and for some $0 < E_1 < E_2 < \infty$ let

$$F(E_1 < H < E_2)\Psi = \Psi.$$

Let $\{f_i\}$ be a smooth decomposition of the identity as given above with (2.3), (2.4). Then

$$\lim_{\tau\to\infty} \|e^{-iH\tau}\Psi - \sum_i f_i(\tfrac{p}{m}) f_i(\tfrac{x}{\tau})\ e^{-iH\tau}\Psi\| = 0. \tag{2.5}$$

Proof of Theorem 2.2. Proposition 2.5(c) below states that

$$\lim_{\tau\to\infty} \|e^{-iH\tau}\Psi - F(E_1<H_0<E_2)\ e^{-iH\tau}\Psi\| = 0. \tag{2.6}$$

By (2.4) this implies

$$\lim_{\tau\to\infty} \| e^{-iH\tau}\Psi - \sum_i f_i^2(p/m)\, e^{-iH\tau}\Psi \| = 0. \tag{2.7}$$

Since the finitely many operators $f_i(p/m)$ are bounded Theorem 2.1 implies (2.5). ■

The support of f in Theorem 2.1 can be chosen very small, so x/τ differs very little from p/m. We have a strong correlation between the position x and the momentum p. For large times τ the scattering state is localized in very special areas of phase space, the latter are "absorbing" for the (interacting) time evolution. This does not contradict the uncertainty relation: the spread of momenta is fixed but the uncertainty of x grows linearly in τ. We prepare the proof of Theorem 2.1 by a technical Lemma. The generator of dilations D is defined as

$$D = \tfrac{1}{2}(x\cdot p + p\cdot x) = x\cdot p - i\nu/2 = p\cdot x + i\nu/2 \tag{2.8}$$

Lemma 2.3. (a) The dense set $\mathfrak{D} = \mathfrak{D}(H_0) \cap \mathfrak{D}(x^2) \subset \mathfrak{D}(D)$ satisfies

$$e^{-iHs}\,\mathfrak{D} = \mathfrak{D}; \tag{2.9}$$

$$\|(1+|x|)^{\mu}\, e^{-iHs}\Psi\| \le \text{const}(\Psi)(1+|s|)^{\mu},\quad \mu = 1,2,\quad \Psi \in \mathfrak{D}; \tag{2.10}$$

$$(H-z)^{-1}\,\mathfrak{D} \subset \mathfrak{D} \text{ for any } z \in \mathbb{C},\ \text{Im } z \ne 0. \tag{2.11}$$

(b) The quadratic form

$$K = i[H,D] - 2H \tag{2.12}$$

which is termwise defined on $\mathfrak{D} \times \mathfrak{D}$ has the property that for any z with Im $z \ne 0$

$$C := (H-z)^{-1}\, K\, (H-z)^{-1} = i[D,(H-z)^{-1}] - 2H(H-z)^{-2} \tag{2.13}$$

extends to a compact operator.

(c) The quadratic form defined termwise on $\mathfrak{D} \times \mathfrak{D}$

$$e^{iHt}\, i[H, \tfrac{m}{2} x^2] e^{-iHt} = e^{iHt}\, D\, e^{-iHt} \tag{2.14}$$

extends to an essentially self-adjoint operator on $\mathfrak{D}$.

Remark 2.4. The compactness in (2.13) remains true if form bounded or highly singular positive potentials are included, see [12]. In particular we note that the commutator

$$i[D, (H-z)^{-1}] \tag{2.15}$$

extends to a bounded operator by (2.13).

Proof. (a) For a proof of (2.9) and (2.10) see e.g [45]. (2.11) follows from (2.10) by representing the resolvent as an exponentially decaying integral of the propagator.

(b) As a quadratic form on $\mathfrak{D} \times \mathfrak{D}$

$$\begin{aligned} K &= i[H_0, D] + i[V_\ell, D] + iV_s D - iDV_s - 2H_0 - 2V \\ &= -x \cdot \nabla V_\ell + iV_s x \cdot p - ip \cdot xV_s + \nu V_s - 2V \end{aligned} \tag{2.16}$$

where we have used that $i[H_0, D] = 2H_0$. By (1.11) $(H_0-z)^{1/2}\, p(H-z)^{-1}$ and its adjoint are bounded. By the assumptions (1.7) and (1.8) on V_ℓ and V_s and the consequences (1.11)-(1.15) all expressions like $x \cdot \nabla V_\ell\, (H-z)^{-1}$, $(H-z)^{-1} V (H-z)^{-1}$, $(H_0-z)^{-1/2}(1+|x|)V_s(H-z)^{-1}$, and their adjoints are compact. This implies compactness of (2.13).

(c) on $\mathfrak{D}$ we have $H = H_0 + V$ as an operator sum. By assumption V is a multiplication operator, thus $[V, x^2] = 0$. Both H_0 and x^2 as well as D are essentially self-adjoint on $C_0^\infty(\mathbb{R}^\nu)$ and they leave this set invariant. Thus the formal calculation $i[H_0, x^2 m/2] = D$ determines the result completely since $C_0^\infty(\mathbb{R}^\nu) \subset \exp(-iHt)\mathfrak{D}$ for all $t \in \mathbb{R}$. ■

Proposition 2.5. (a) Let $g \in C_0^\infty(\mathbb{R})$ and $\Psi \in \mathcal{H}^{cont}(H)$, then

$$\lim_{\tau\to\infty} \|\{g(D/\tau)-g(2H)\}\, e^{-iH\tau}\Psi\| = 0. \tag{2.17}$$

(b) For any $\Psi \in \mathcal{H}^{cont}(H)$

$$\operatorname*{w-lim}_{|\tau|\to\infty} e^{-iH\tau}\,\Psi = 0. \tag{2.18}$$

(c) Let $\Psi \in \mathcal{H}^{cont}(H)$ satisfy $\Psi = F(E_1<H<E_2)\Psi$, then

$$\lim_{\tau\to\infty} \|e^{-iH\tau}\Psi - F(E_1<H_0<E_2)\, e^{-iH\tau}\Psi\| = 0. \tag{2.19}$$

Proof. (a) Since $\mathcal{D}' = (H-z)^{-1}\,\mathcal{D}$ is dense we can find a $\Psi' \in \mathcal{D}'$ such that for given $\varepsilon > 0$

$$\|\Psi-\Psi'\| < \varepsilon. \tag{2.20}$$

Then the error in (2.17) from replacing Ψ by Ψ' is bounded by $\varepsilon \cdot 2\|g\|_\infty$ uniformly in τ. Expanding the functions of operators as Fourier integrals one obtains

$$\|\{g(D/\tau)-g(2H)\}\, e^{-iH\tau}\Psi'\| \tag{2.21}$$

$$\le (2\pi)^{-1/2}\int ds(1+|s|)|\hat{g}(s)| \cdot \sup_{|\sigma|\le S} \frac{1}{1+|\sigma|}\|[e^{i\sigma D/\tau}-e^{i2H\sigma}]\, e^{-iH\tau}\Psi'\|$$

$$+\ 2\|\Psi'\|\ (2\pi)^{-1/2} \int_{|s|>S} ds\ (1+|s|)|\hat{g}(s)|$$

$$\le \varepsilon + \text{const}(g) \sup_{|s|\le S} \|[D/\tau-2H]\, e^{i2Hs}\, e^{-iH\tau}\,\Psi'\|$$

for suitably chosen $S = S(\varepsilon,g)$. In the last step we have used the Duhamel formula

$$e^{i\sigma A} - e^{i\sigma B} = i\int_0^\sigma ds\ e^{iA(\sigma-s)}(A-B)\, e^{iBs}. \tag{2.22}$$

By Remark 2.4 the following commutator term

$$(2.23)\quad \|\tau^{-1}De^{-iH(\tau-2s)}\Psi' - \tau^{-1}(H-z)^{-1}D(H-z)^{-1}e^{-iH(\tau-2s)}(H-z)^{2}\Psi'\|$$

$$\le |\tau|^{-1}\|[D,(H-z)^{-1}]\|\cdot\|(H-z)\Psi'\|$$

is smaller than ε for all $|\tau| \geqslant T(\varepsilon)$ uniformly in s. The following calculation is justified as an equality for quadratic forms on $\mathfrak{D} \times \mathfrak{D}$, by Lemma 2.3b) it extends to an operator identity:

$$(2.24)\quad \tau^{-1}\, e^{iH\tau}(H-z)^{-1}\, D(H-z)^{-1}\, e^{-iH\tau}$$

$$= \tau^{-1}(H-z)^{-1}\, D(H-z)^{-1}$$

$$+ \frac{1}{\tau}\int_0^\tau dt\; e^{iHt}(H-z)^{-1}\; i[H,D](H-z)^{-1}\; e^{-iHt}$$

$$= \tau^{-1}(H-z)^{-1}\, D(H-z)^{-1} + 2H(H-z)^{-2} + \frac{1}{\tau}\int_0^\tau dt\; e^{iHt}\; C\; e^{-iHt}.$$

Thus for $|\tau| \geqslant T(\varepsilon)$

$$(2.25)\quad \sup_{|s|\le S} \|[D/\tau-2H]\; e^{-iH(\tau-2s)}\;\Psi'\|$$

$$\le \varepsilon + |\tau|^{-1} \sup_{|s|\le S} \|(H-z)^{-1}\; D\; e^{-iH2s}(H-z)\Psi'\| +$$

$$+ \|\frac{1}{\tau}\int_0^\tau dt\; e^{iHt}\; C\; e^{-iHt}\; P^{cont}(H)\|\cdot\|(H-z)^{2}\;\Psi'\| +$$

$$+ \|C\|\cdot\|(H-z)^{2}\; P^{pp}(H)\;\Psi'\|.$$

The norm in the second term

$$(2.26)\quad \|(H-z)^{-1}[p\cdot x + i\nu/2]\; e^{iH2s}(H-z)\Psi'\|$$

$$\le \|(H-z)^{-1}|p|\| \cdot \||x|e^{iH2s}(H-z)\Psi'\| + \nu/2$$

is uniformly bounded on compact intervals for the parameter s by (2.10). A standard estimate used in ergodic theory shows that for any compact operator C and any self-adjoint generator H

(2.27) $$\lim_{\tau\to\infty} \left\|\frac{1}{\tau}\int_0^\tau dt\, e^{iHt}\, C\, e^{-iHt}\, P^{cont}(H)\right\| = 0.$$

For a proof see Appendix to XI.17 in [46] or Lemma 4.2 in [9] for exactly this statement. For the last summand in (2.25) observe that typically $H\, P^{pp}(H)$ is bounded and

$$\|P^{pp}(H)\Psi'\| = \|P^{pp}(H)(\Psi-\Psi')\| < \varepsilon.$$

In general one can give explicitly an approximate Ψ' which satisfies both (2.20) and

(2.28) $$\|(H-z)^2\, P^{pp}(H)\Psi'\| < \varepsilon.$$

(b) Since $\Psi \in \mathcal{H}^{cont}(H)$ there is for any $\varepsilon > 0$ a real $g \in C_0^\infty(\mathbb{R})$, $0 \notin \operatorname{supp} g$, with

$$\|[\mathbb{1}-g(2H)]\,\Psi\| < \varepsilon.$$

For any such g and any self-adjoint D

(2.29) $$\operatorname*{s-lim}_{|\tau|\to\infty} g(D/\tau) = 0.$$

Thus for large enough $|\tau|$ using (a)

(2.30) $$|(\Phi, e^{-iH\tau}\Psi)| \le$$

$$\le \|\Phi\|\,(\varepsilon+\|[g(D/\tau)-g(2H)]\, e^{-iH\tau}\Psi\|) + \|g(D/\tau)\Phi\|\cdot\|\Psi\|$$

$$< 2\varepsilon\|\Phi\|.$$

(c) With (b) and the compactness (1.14) we obtain strong resolvent convergence of the kinetic energy to the full energy on $\mathcal{H}^{cont}(H)$. With H_0 and $H\restriction\mathcal{H}^{cont}(H)$ both being purely continuous operators one can admit bounded discontinuous functions of H and H_0, i.e. for any

$\Psi \in \mathcal{H}^{cont}(H)$

$$\lim_{\tau\to\infty} \|[F(E_1 < H < E_2) - F(E_1 < H_0 < E_2)]\, e^{-iH\tau}\Psi\| = 0. \qquad \blacksquare$$

Lemma 2.6. Let $f \in C_0^\infty(\mathbb{R}^\nu)$ and let $\Phi \in \mathfrak{D}(x) \cap \mathfrak{D}(p)$. Then there is a constant $C(f)$ such that for the self-adjoint operators x and p

$$\|[f(\tfrac{x}{\tau}) - f(\tfrac{p}{m})]\,\Phi\| \le C(f)\{\|[\tfrac{x}{\tau} - \tfrac{p}{m}]\,\Phi\| + \tfrac{1}{2m\tau}\|\Phi\|\}.$$

Proof. By the Baker-Campbell-Hausdorff formula

$$(2\pi)^{\nu/2}\,[f(\tfrac{x}{\tau}) - f(\tfrac{p}{m})]$$

$$= \int d^\nu q\; \hat{f}(q)[e^{iq\cdot x/\tau} - e^{iq\cdot p/m}]$$

$$= \int d^\nu q\; \hat{f}(q)e^{iq\cdot p/m}[\exp\{iq\cdot(\tfrac{x}{\tau} - \tfrac{p}{m} - \tfrac{q}{2mt})\} - \mathbb{1}].$$

$$\|[\exp\{iq\cdot(\tfrac{x}{\tau} - \tfrac{p}{m} - \tfrac{q}{2mt})\} - \mathbb{1}]\,\Phi\|$$

$$\le |q|\;\|(\tfrac{x}{\tau} - \tfrac{p}{m})\,\Phi\| + \frac{|q|^2}{2mt}\|\Phi\|.$$

Thus

$$\|[f(\tfrac{x}{\tau}) - f(\tfrac{p}{m})]\,\Phi\|$$

$$\le (2\pi)^{-\nu/2}\int d^\nu q\;(1+|q|^2)|\hat{f}(q)|\;\{\|(\tfrac{x}{\tau} - \tfrac{p}{m})\,\Phi\| + \tfrac{1}{2m\tau}\|\Phi\|\}. \qquad \blacksquare$$

Proof of Theorem 2.1. Similarly to the proof of Proposition 2.5 (a) it is sufficient to find for any $\Psi \in \mathcal{H}^{cont}(H)$, $\varepsilon > 0$ a Ψ', $\|\Psi - \Psi'\| < \varepsilon$, such that

$$\limsup_{\tau\to\infty} \|[\tfrac{x}{\tau} - \tfrac{p}{m}]\, e^{-iH\tau}\Psi'\|^2 < \text{const}\ \varepsilon. \tag{2.31}$$

The squared norm in (2.31) is proportional to

(2.32) $$(\Psi', e^{iH\tau}\{\frac{m}{2}(\frac{x}{\tau})^2 - \frac{D}{\tau} + H_0\}e^{-iH\tau}\Psi').$$

For any $\Psi' \in \mathfrak{D}$

(2.33) $$\tau^{-2} (\Psi', e^{iH\tau} \frac{m}{2} x^2 e^{-iH\tau}\Psi')$$

$$= \tau^{-2} (\Psi', \frac{m}{2} x^2 \Psi')$$

$$+ \frac{1}{\tau^2} \int_0^\tau dt \ (\Psi', e^{iHt} i[H, \frac{m}{2} x^2] e^{-iHt}\Psi').$$

The first summand vanishes as $\tau \to \infty$. Using (2.14) the second can be rewritten as

(2.34) $$\frac{1}{\tau^2} \int_0^\tau dt \cdot t \cdot (\Psi', e^{iHt} \{\frac{D}{t} - 2H\} e^{-iHt}\Psi') + (\Psi', H \Psi').$$

For Ψ' as chosen in the proof of Proposition 2.5 we apply the estimate (2.25) twice to obtain

(2.35) $$(\Psi', e^{iH\tau}\{\frac{m}{2}(\frac{x}{\tau})^2 - \frac{D}{\tau} + H\} e^{-iH\tau}\Psi') < \text{const } \varepsilon.$$

Finally it remains to estimate

(2.36) $$|(\Psi', e^{iH\tau}[H - H_0] e^{-iH\tau}\Psi')|$$

$$\leq \|(H_0+z)\Psi'\| \cdot \|(H_0-z)^{-1} V(H-z)^{-1} e^{-iH\tau} P^{cont}(H)(H-z)\Psi'\|$$

$$+ \|\Psi'\| \cdot \|V(H-z)^{-1}\| \cdot \|(H-z)P^{pp}(H)\Psi'\|.$$

By Proposition 2.5(b) and compactness (1.14) of $(H_0-z)^{-1} V(H-z)^{-1}$ the first summand vanishes as $|\tau| \to \infty$. The second is bounded by const$\cdot\varepsilon$ as discussed above. ∎

A proof of the results in this section under weaker assumptions on the potentials is given in [12]. It is sufficient that ∇V_ℓ is of short range, the faster decay in (1.8) will be needed only in later sections. Closely related results are given by Sinha and Muthuramalingam in [49].

III. Completeness for Short-Range Potentials.

In this section we control the future time evolution for states in the range of $f(p/m)\, f(x/\tau)$ for functions f with the property (2.3) if only short range forces are present. As an application we show asymptotic completeness. The inclusion of long-range forces follows in the next sections.

We start with a well known estimate saying that the free quantum time evolution mainly propagates into the directions where a classical free particle with the same phase-space localization would travel. The "quantum tails" in the "classically forbidden" region have rapid decay.

Proposition 3.1. Let $g \in C_0^\infty(\mathbb{R}^\nu)$ have

$$(3.1) \qquad \operatorname{supp} g \subset \{w \in \mathbb{R}^\nu \mid |w-v| < u\}$$

for some $v \in \mathbb{R}^\nu$, $u > 0$. Let M, M' be measurable subsets of $\mathbb{R}^\nu$ with $r := \operatorname{dist}\{M', M+vt\} - u|t| \geq 0$. Then for any $n \in \mathbb{N}$

$$(3.2) \qquad \| F(x\epsilon M')\, e^{-iH_0 t}\, g(p/m)\, F(x\epsilon M) \| \leq C_n (1+|t|+r)^{-n},$$

for constants C_n which depend only on g. Moreover, if (3.1) holds for a parametrized family of functions which depend continuously in $C^\infty(\mathbb{R}^\nu)$ on the parameters, then the constants C_n can be chosen uniform on compact parameter sets.

Sketch of the proof. For a detailed discussion see e.g. Section II of [11]. By a Galilei transformation one can set $v = 0$ without loss of generality. It is sufficient to show that the Fourier transform $\hat{\varphi}_t(y)$ of

$$\varphi_t(p) = \exp(-itp^2/2m)\ g(p/m)$$

has rapid decay in $|y| + |t|$ for $|y| > u\ |t|$. By (3.1) there is a small $a > 0$ such that

$$\text{supp } g \subset \{w|\ |w| < u - 2a\}.$$

This is true for a compact family of functions as well. Then the phase function in the Fourier integral of φ_t is

$$p\cdot y + t\ p^2/2m = (|y|+u|t|)\{[p\cdot y+t\ p^2/2m]/(|y|+u|t|)\}.$$

The gradient with respect to p of the expression in braces is

$$|[y - t\ p/m]/(|y|+u|t|)| \geqslant a/u > 0,$$

where the inequality holds if $|y| > u|t|$ and $(p/m) \in \text{supp } g$. Thus there are no points of stationary phase and the standard estimates (see e.g. Appendix 1 to XI.3 in [46]) show rapid decay in $|y| + |t|$. ■

Corollary 3.2. Let $f \in C_0^\infty(\mathbb{R}^\nu)$ satisfy (3.1) with $u \leq |v|/2$. Then for $t \cdot \tau > 0$, $b \in \mathbb{Z}$

$$\|F(|x|<(t+\tau)|v|/2)\ e^{-iH_0t}\ (H_0-z)^b\ f(p/m)\ f(x/\tau)\| \tag{3.3}$$

$$\leq C_n(1+|t+\tau|)^{-n}.$$

Proof. Since supp f is compact there is an $a > 0$ such that for $\tau, t \geqslant 0$

$$\text{supp } f(x/\tau) + vt \subset \{x \in \mathbb{R}^\nu|\ |x| \geqslant \tau|v|/2 + \tau a + |v|t\}.$$

Thus

$$\mathrm{dist}(\{x|\ |x|<(t+\tau)|v|/2\}, \mathrm{supp}\, f(x/\tau)+vt) - t|v|/2$$

$$= r \geq \tau \cdot a \geq 0.$$

With f also $[(p^2/2m)-z]^b\, f(p/m)$ satisfies (3.1) and therefore (3.2) implies (3.3). Similarly for $t,\tau < 0$. ■

Proposition 3.3. Let $H = H_0 + V_s$ where V_s satisfies (1.6), (1.7). Then for any $f \in C_0^\infty(\mathbb{R}^\nu)$ which satisfies (3.1) with $u \leq |v|/2$

$$\lim_{\tau\to\infty} \sup_{t\geq 0} \|[e^{-iHt} - e^{-iH_0 t}]\, f(p/m) f(x/\tau)\| = 0. \tag{3.4}$$

Proof. By "Cook's method" uniformly in $t \geq 0$

$$\|[\mathbb{1}-e^{iHt}\, e^{-iH_0 t}]\, f(p/m)\, f(x/\tau)\| \tag{3.5}$$

$$\leq \int_0^\infty ds \|\tfrac{d}{ds}\, e^{iHs}\, e^{-iH_0 s}\, f(p/m) f(x/\tau)\|$$

$$\leq \int_0^\infty ds \|V_s (H_0-z)^{-1}\, e^{-iH_0 s} (H_0-z) f(p/m)\, f(x/\tau)\|$$

$$\leq \int_0^\infty ds \|V_s (H_0-z)^{-1} F(|x|>(\tau+s)|v|/2)\| \cdot \|(H_0-z) f(p/m) f(x/\tau)\|$$

$$+ \|V_s(H_0-z)^{-1}\| \int_0^\infty ds \|F(|x|<(\tau+s)|v|/2)\, e^{-iH_0 s} (H_0-z) f(p/m) f(x/\tau)\|.$$

By the short-range condition (1.7′) on the potential the first integral converges and it tends to zero as $\tau \to \infty$. The second integrand has rapid decay in $(\tau+s)$ by Corollary 3.2, thus its integral decays faster than any inverse power of τ. ■

Now it is easy to complete the proof of existence and completeness of the wave operators for short-range potentials.

Theorem 3.4. Let $H = H_0 + V_s$ where V_s satisfies (1.6), (1.7) and let H' denote H or H_0. Then for any $\Psi \in \mathcal{H}^{\mathrm{cont}}(H')$

$$(3.6) \qquad \lim_{\tau\to\infty} \sup_{t\geq 0} \| [e^{-iHt} - e^{-iH_0 t}] \, e^{-iH'\tau} \Psi \| = 0 .$$

Proof. The vectors $\Psi \in \mathcal{H}^{cont}(H')$ for which there are $0 < E_1 < E_2 < \infty$ such that $\Psi = F(E_1 < H' < E_2)\,\Psi$ form a dense set in $\mathcal{H}^{cont}(H')$. It is sufficient to verify (3.6) for these vectors. Theorem 2.2 holds for any H considered, in particular for H_0 as well. Therefore

$$(3.7) \quad \lim_{\tau\to\infty} \sup_{t\geq 0} \| [e^{-iHt} - e^{-iH_0 t}] \, \{\mathbb{1} - \sum_i f_i(p/m) f_i(x/\tau)\} \, e^{-iH'\tau} \Psi \| = 0 .$$

To each of the finitely many summands labelled by i Proposition 3.3 applies which completes the proof. ■

Corollary 3.5. Under the assumptions of Theorem 3.4 the wave operators

$$(3.8) \qquad \Omega_\pm = s\text{-}\lim_{t\to\pm\infty} e^{iHt} \, e^{-iH_0 t}$$

exist and are complete, i.e.

$$(3.9) \qquad \mathrm{Ran}\, \Omega_\pm = \mathcal{H}^{cont}(H) = \mathcal{H}^{ac}(H),$$

in particular H has no singular continuous spectrum.

Proof. Let $H' = H_0$ and $\Psi \in \mathcal{H} = \mathcal{H}^{cont}(H_0)$.

$$(3.10) \qquad \| e^{iH(t+\tau)} \, e^{-iH_0(t+\tau)} \Psi - e^{iH\tau} \, e^{-iH_0\tau} \Psi \|$$

$$= \| [e^{-iH_0 t} - e^{-iHt}] \, e^{-iH_0\tau} \Psi \| .$$

Thus (3.6) is the Cauchy convergence criterion in this case. It is well known that $\Psi \in \mathrm{Ran}\, \Omega_\pm$ iff

$$\lim_{t\to\pm\infty} e^{iH_0 t} \, e^{-iHt} \, \Psi \text{ exists,}$$

i.e., if (3.6) holds with $H' = H$. By the intertwining property one automatically has Ran $\Omega_{\pm} \subset \mathcal{H}^{ac}(H)$ which implies (3.9). ■

IV. An Approximate Time Evolution for Long-Range Potentials

If long-range forces like the physically important Coulomb potential are present the free time evolution is no longer a good approximation of the assymptotic time evolution of a scattering state. The modified free time evolution as given by Dollard [5] works for the class of potentials considered here. If the decay is even slower then a better approximation has to be used, see e.g. [22,23], [27]. The effect is purely classical, only the estimates are slightly more involved in quantum mechanics since the position and momentum operators do not commute.

In this section we construct an "intermediate time evolution" which is a good approximation of the true time evolution on the ranges of the operators $f_i(p/m)f_i(x/\tau)$. The propagation properties of a state are much easier to control than for the fully interacting time evolution. This will be used in the next section to show that the even simpler modified free time evolution of Dollard is a good approximation for the further future.

We choose a sequence of quickly increasing times

(4.1) $$\tau_k := k^{2\rho}, \qquad t_k := \tau_{k+1} - \tau_k \sim k^{2\rho-1};$$

where

(4.2) $$1 < \rho < 1/(1-2\delta).$$

The choice of ρ will become clear later. A scattering state with strictly positive energy leaves the origin with some minimal speed. Correspondingly we use only the "tail part" of the long-range potential in our approximation. Let $\varphi \in C_0^\infty(\mathbb{R}^\nu)$ satisfy $0 \le \varphi(x) \le 1$,

(4.3) $$\varphi(x) = 1 \text{ for } |x| \leq 1/2, \quad \varphi(x) = 0 \text{ for } |x| \geq 1.$$

Then we set for some $u_0 > 0$ (to be determined later)

(4.3′) $$V_k(x) := V_\ell(x)\cdot[1-\varphi(x/u_0\tau_k)], \quad k \in \mathbb{N}.$$

Observe that k is a running index while the subscript ℓ denotes the long-range part! According to the property (1.9) of the long-range potential V_ℓ the family of cutoff potentials V_k satisfies

(4.4) $$\sup|(\nabla V_k)(x)| \leq \text{const.}(\tau_k)^{-(\delta+3/2)},$$

(4.5) $$\sup|(\Delta V_k)(x)| \leq \text{const.}(\tau_k)^{-(\delta+2)}.$$

Let $f \in C_0^\infty(\mathbb{R}^\nu)$ satisfy for some $0 \neq v \in \mathbb{R}^\nu$

(4.6) $$\text{supp } f \subset \{w \in \mathbb{R}^\nu | \; |w-v| < |v|/2\}.$$

For this v let $\bar{f}, \bar{g} \in C_0^\infty(\mathbb{R}^\nu)$ satisfy

(4.7) $$\bar{g}(w) = 1 \text{ if } |w-v| \leq |v|/2,$$

$$\text{supp } \bar{g} \subset \{w \in \mathbb{R}^\nu | |w-v| < 2|v|/3\},$$

(4.8) $$\bar{f}(w) = 1 \text{ if } |w-v| \leq 2|v|/3, \quad \text{and}$$

$$\bar{f}(w) = 0 \text{ if } |w-v| \geq 3|v|/4.$$

For our purpose here it is sufficient to define the intermediate time evolution for times from the particular sequence $\{\tau_k\}$.

Definition 4.1. For a given f which satisfies (4.6) pick functions $\bar{f}, \bar{g}$ and the family V_k of cutoff potentials with $0 < u_0 < |v|/3$. Then for $n > m$

$$(4.9)\qquad U(\tau_n,\tau_m) := \bar{g}(p/m)\bar{f}(x/\tau_n)\, e^{-iV_{n-1}t_{n-1}}\, e^{-iH_0 t_{n-1}} \times\cdots\times$$

$$\times\, e^{-iV_{m+1}t_{m+1}}\, e^{-iH_0 t_{m+1}}\, e^{-iV_m t_m}\, e^{-iH_0 t_m}.$$

Note that the times are ordered to increase from right to left. See also Remarks 4.3 below. We show that U is a good approximation of the fully interacting time evolution on certain states.

<u>Proposition</u> <u>4.2</u>. Let H be as defined in Section I and let $f \in C_0^\infty(\mathbb{R}^\nu)$ satisfy (4.6). For U as defined in (4.9)

$$(4.10)\qquad \lim_{m\to\infty} \sup_{n>m} \| [e^{-iH(\tau_n-\tau_m)} - U(\tau_n,\tau_m)]\, f(p/m)f(x/\tau_m)\| = 0.$$

<u>Proof</u>. We have to construct an auxiliary sequence $f_k \in C_0^\infty(\mathbb{R}^\nu)$, $k \in \mathbb{N}$. Let χ_k denote the characteristic function of a ball around v with radius

$$(4.11)\qquad \frac{|v|}{2} + \frac{2}{c}\sum_{j=1}^{k} j^{-\rho}.$$

With $\psi \in C_0^\infty(\mathbb{R}^\nu)$, $\psi(q) \geq 0$, $\psi(q) = 0$ for $|q| \geq 1$ and

$$\int d^\nu q\, \psi(q) = 1 \quad \text{let}$$

$$(4.12)\qquad f_k := c(k+1)^\rho\, \chi_k * \psi[c(k+1)^\rho \cdot].$$

Then $0 \leq f_k \leq 1$, $f_k \in C_0^\infty(\mathbb{R}^\nu)$,

$$(4.13)\qquad f_1(w) = 1 \quad \text{if} \quad |w-v| \leq |v|/2,$$

$$(4.14)\qquad f_k(w) = 1 \quad \text{if} \quad w \in \operatorname{supp} f_{k-1},$$

and since $\rho > 1$ for sufficiently large c:

$$(4.15)\qquad f_k(w) = 0 \quad \text{if } |w-v| \geq 2|v|/3 \text{ for all } k.$$

Moreover

$$\sup_w |(\nabla f_k)(w)| \le \text{const. } k^{\rho}. \tag{4.16}$$

By (4.11), (4.14) we have in x-space

$$\operatorname{dist}(\operatorname{supp} f_k(x/\tau_k) + vt_k,\ \operatorname{supp}[1-f_{k+1}(x/\tau_{k+1})]) \geq t_k|v|/2. \tag{4.17}$$

For this family of cutoff functions we have

$$\bar{g}\cdot f = f, \tag{4.18}$$

$$f_k\cdot f = f, \tag{4.19}$$

$$\bar{f}\cdot f_k = f_k \text{ for all } k. \tag{4.20}$$

With (4.6), (4.17) and Proposition 3.1 we get

$$\|[\mathbb{1}-f_{k+1}(x/\tau_{k+1})]\ e^{-iH_0t_k}\ f(p/m)f_k(x/\tau_k)\| \le C_n(1+t_k)^{-n}, \tag{4.21}$$

and for any $u_0 < |v|/3$ and $h \in C^{\infty}(\mathbb{R}^{\nu})$

$$\sup_{t\geq 0} \|F(|x| \le u_0\tau_k)\ e^{-iH_0t}\ h(p)\ f(p/m)\ f_k(x/\tau_k)\| \le c_n(1+\tau_k)^{-n}. \tag{4.22}$$

Moreover it is easy to verify that for any $g \in C_0^{\infty}(\mathbb{R}^{\nu})$

$$\|[g(p), f_k(x/\tau_k)]\| \le \text{const } k^{\rho}/\tau_k = \text{const } k^{-\rho}. \tag{4.23}$$

We will encounter (4.21)-(4.23) as error terms below. They are all summable in k! To get a family with (4.14) and (4.15) the power of k in (4.16) has to be greater than one. Summability of (4.23) requires that τ_k grows faster than the square of k. This explains the choice of the lower bound $\rho > 1$ in (4.2).

We use the standard convention that empty products are $\mathbb{1}$ and by Π' we denote the time-ordered product with increasing times from right to left. The shorthand f_k stands for $f_k(x/\tau_k)$ and $f \equiv f(p/m)$. By induction one easily verifies for $n \geq m + 1$

$$(4.24) \quad \prod_{k=m}^{n-1}{}' [e^{-iV_k t_k} e^{-iH_0 t_k}] f f_m$$

$$= \sum_{j=m}^{n-1} \prod_{k=j+1}^{n-1}{}' [e^{-iV_k t_k} e^{-iH_0 t_k}] A_j \prod_{r=m}^{j-1}{}' [e^{-iV_r t_r} e^{-iH_0 t_r} f_r]$$

$$+ f f_n \prod_{r=m}^{n-1}{}' [e^{-iV_r t_r} e^{-iH_0 t_r} f_r],$$

where

$$(4.25) \quad A_j = e^{-iV_j t_j} e^{-iH_0 t_j} f f_j - f f_{j+1} e^{-iV_j t_j} e^{-iH_0 t_j} f_j .$$

$$(4.26) \quad \|A_j\| \leq \|(\mathbb{1}-f_{j+1}) e^{-iH_0 t_j} f f_j\|$$

$$+ \|[f_{j+1}, f\| + \|[e^{-iV_j t_j}, f]\| .$$

By (4.21) and (4.23) the first two terms are summable in j. The same is true for the last summand by (4.4) since

$$(4.27) \quad \|[e^{-iV_j t_j}, f(p)]\| \leq \text{const } t_j \|\nabla V_j\| \leq \text{const } t_j/(\tau_j)^{\delta+3/2}$$

$$\leq \text{const}/\tau_j^{(1+\varepsilon)}, \; \varepsilon > 0.$$

With (4.18)-(4.20), and (4.23) clearly

$$(4.28) \quad f(p/m)\, f(x/\tau_m) = f(p/m)\, f_m(x/\tau_m)\, f(x/\tau_m),$$

(4.29) $$\lim_{n\to\infty} \| [1\!\!1 - \bar{g}(p/m)\bar{f}(x/\tau_m)]\ g(p/m) f_n(x/\tau_n) \| = 0.$$

Thus we have shown

(4.30) $$\lim_{m\to\infty} \sup_{n>m} \| \{ U(\tau_n, \tau_m)\ f(p/m)\ f(x/\tau_m) -$$

$$- f(p/m) f_n \prod_{k=m}^{n-1}{}' [e^{-iV_k t_k} e^{-iH_0 t_k} f_k] f(x/\tau_m) \| = 0.$$

Similarly one shows by induction

(4.31) $$\prod_{k=m}^{n-1} e^{-iHt_k} f\ f_m$$

$$= f\ f_n \prod_{k=m}^{n-1}{}' [e^{-iV_k t_k} e^{-iH_0 t_k} f_k]$$

$$+ \sum_{j=m}^{n-1} \prod_{k=j+1}^{n-1} e^{-iHt_k} B_j \prod_{r=m}^{j-1}{}' [e^{-iV_r t_r} e^{-iH_0 t_r} f_r],$$

where

(4.32) $$B_j = \{ e^{-iHt_j} f - f\ f_{j+1} e^{-iV_j t_j} e^{-iH_0 t_j} \} f_j$$

$$= \{ e^{-iHt_j} - e^{-iV_j t_j} e^{-iH_0 t_j} \} f\ f_j + A_j .$$

(4.33) $$\| B_j - A_j \| \le$$

$$\le \int_0^{t_j} dt\ \| \frac{d}{dt} e^{iHt} e^{-iV_j t} e^{-iH_0 t} f\ f_j \|$$

$$\le \int_0^{t_j} dt \{ \| V_s (H_0 - z)^{-1} e^{-iH_0 t} (H_0 - z) f\ f_j \|$$

$$+ \| (V_\ell - V_j) e^{-iH_0 t} f\ f_j \| + \| [H_0, e^{-iV_j t}]\ f(p) \| \}.$$

By the short range property (1.7) of V_s

$$t_j \cdot \|V_s(H_0-z)^{-1} F(|x|>u_0\tau_j)\| \tag{4.34}$$

is summable in j. The estimate (4.22) gives rapid decay in j for

$$\|F(|x| < u_0\tau_j)\ e^{-iH_0t}\ (H_0-z)^{\beta}\ f\ f_j\|.$$

Since $V_\ell(x) - V_j(x) = 0$ for $|x| > u_0\tau_j$ this is sufficient for the second term as well. The integral of the third term is bounded by a constant times

$$t_j\{\|\nabla V_j\|\cdot t_j\cdot\|p\ g(p)\| + (\|\nabla V_j\|\cdot t_j)^2 + \|\Delta V_j\|\cdot t_j)\}. \tag{4.35}$$

By (4.1), (4.4), and (4.5) this is summable in j if $\rho < 1/(1-2\delta)$. Here the upper bound (4.2) on ρ comes in.

We have shown summability in j of $\|B_j\|$. With (4.30) this completes the proof of Proposition 4.2. ■

Remarks 4.3. (a) it is clear from the proof that the product $\bar{g}(p/m)\bar{f}(x/\tau_n)$ could have been omitted from the definition (4.9) of U. We have included it for later convenience. Moreover one could have used V_ℓ everywhere instead of V_k because of (4.21) and $[V_\ell(x)-V_k(x)]\ f_k(x/\tau_k) = 0$. This approximate time evolution (which was given originally in Section VI of [11]) is independent of f and it satisfies Proposition 4.2 as well. The propagation properties in phase space, however, apparently are easier to control for U as defined in (4.9).

(b) The exponent $V_k t_k$ is bounded by k raised to the power $\rho(1-2\delta) - 1 < 0$. Thus one can replace the exponential by its Taylor polynomial up to a summable error. If e.g. $\delta > 1/4$ (which includes Coulomb forces) and if one chooses $1 < \rho < (2-4\delta)^{-1}$ then $\exp(-iV_k t_k)$ can be replaced by $(\mathbb{1}-iV_k t_k)$.

(c) These and related approximate time evolutions have been

treated in detail by M. Schneider [47] and M. Knick [28], see also [15]. In particular much more general regions of space can be treated like, e.g., half spaces. Another geometrical configuration is studied in Section VI. These time evolutions are in a certain sense quasiclassical approximations. In U functions of x and functions of p are separated for intervals of increasing length t_k. The upper bound t_k comes from the commutator terms (4.35) which have to be small. On the other hand for long enough times the free quantum time evolution is well approximated by the free classical one, resulting in the construction of a family f_k which satisfies (4.21). Here the lower bound on t_k enters. A better quasiclassical approximation using Fourier integral operators allows to treat a much larger class of potentials as shown by Kitada and Yajima [27]. However, their treatment is technically very demanding in contrast to our elementary estimates.

V. Completeness for Long-Range Potentials.

In this section we prove asymptotic completeness for Hamiltonians with long-range potentials, Theorem 5.5. Our first goal is to show propagation properties in phase space for the approximate time evolution U. Beyond the results of Section II we show a certain decay rate in τ which is best possible. For the free time evolution an easy calculation shows

$$(5.1) \qquad [x-tp/m]\, e^{-iH_0 t} = e^{-iH_0 t}\, x .$$

Thus for $\Psi \in \mathfrak{D}$

$$(5.2) \qquad \left(\frac{x}{t} - \frac{p}{m}\right) e^{-iH_0 t}\, \Psi = \frac{1}{t}\, e^{-iH_0 t}\, x\, \Psi \sim \frac{1}{t}.$$

If long range forces are present this is no longer true. Coulomb forces cause an $(\ln t)/t$ behaviour and potentials with a decay like (1.8), $\delta < 1/2$ may induce a decay as slow as $t^{-(\delta+1/2)}$. This can be seen from estimates of classical trajectories.

Proposition 5.1. Let V_ℓ satisfy (1.8) with $0 < \delta \le 1/2$ and let f and U be given as in Definition 4.1. Then for $\tau_n \geq \tau_m$

$$\|(\frac{x}{\tau_n} - \frac{p}{m})\, U(\tau_n,\tau_m)\, f(p/m)f(x/\tau_m)\| \tag{5.3}$$

$$\le \begin{cases} \mathrm{const}(\tau_m) \cdot \tau_n^{-(\delta+1/2)}, & 0 < \delta < \frac{1}{2}, \\ \mathrm{const}[\tau_m + \ell n(\tau_n/\tau_m)]/\tau_n, & \delta = \frac{1}{2}. \end{cases}$$

Proof. Clearly

$$[(\frac{x}{\tau_n} - \frac{p}{m}), \bar{g}(p/m)\bar{f}(x/\tau_n)] \sim 1/\tau_n \tag{5.4}$$

satisfies the desired estimate. By induction one easily verifies using (5.1)

$$(mx - p\tau_n) \prod_{k=m}^{n-1}{}' e^{-iV_k t_k} e^{-iH_0 t_k} \tag{5.5}$$

$$= \prod_{k=m}^{n-1}{}' e^{-iV_k t_k} e^{-iH_0 t_k} (mx - p\tau_m)$$

$$- \sum_{j=m}^{n-1} \prod_{k=j+1}^{n-1}{}' \ldots \tau_{j+1}[p, e^{-iV_j t_j}]\, e^{-iH_0 t_j} \prod_{r=m}^{j-1} \ldots\,.$$

With

$$\|\tau_{j+1}[p, e^{-iV_j t_j}]\| \sim j^{-1+\rho(1-2\delta)} \tag{5.6}$$

the sum over j is bounded by

$$\mathrm{const.}\ \tau_n^{\frac{1}{2}-\delta}\ (\mathrm{resp.}, \ell n(\tau_n/\tau_m).) \tag{5.7}$$

The contribution from the first term in (5.5) gives

$$\|(mx - p\tau_m)\, f(p/m)f(x/\tau_m)\| \le \mathrm{const.}\ \tau_m. \tag{5.8}$$

Summing up the estimates gives

(5.9) $$\|(\frac{x}{\tau_n} - \frac{p}{m})\ U(\tau_n,\tau_m)\ f(p/m)f(x/\tau_m)\|$$

$$\le \begin{cases} \text{const.}(1+\tau_m+\tau_n^{\frac{1}{2}-\delta})/\tau_n, & 0 < \delta < \frac{1}{2}, \\ \text{const}[\tau_m+\ell n(\tau_n/\tau_m)]/\tau_n, & \delta = \frac{1}{2}. \end{cases}$$

This implies (5.3). ■

Dollard's modified free time evolution U_D is defined as [5]

(5.10) $$U_D(T,\tau) := e^{-iH_0(T-\tau)}\ U'(T,\tau),$$

(5.11) $$U'(T,\tau) := \exp(-i \int_\tau^T dt\ V_\ell(tp/m)).$$

The "quantum tail" propagating into the classically frobidden region is very small again.

<u>Lemma</u> <u>5.2</u>. Let V_ℓ satisfy (1.9) and $\bar{f},\bar{g}$ as given in (4.7),(4.8). Then for any $u_0 < |v|/4$, b = 0 or 1:

(5.12) $$\|F(|x| \le u_0T)\ U_D(T,\tau)\ (H_0-z)^b\ \bar{g}(p/m)\bar{f}(x/\tau)\|$$

$$\le c_n(1 + T)^{-n}$$

uniformly in $1 \le \tau \le T$.

For a proof see Corollary 2.12 in [11]. It is a simple extension of Proposition 3.1. We extend also Proposition 5.1. to include Dollard's propagator:

<u>Lemma</u> <u>5.3</u>. Let f,U, and U_D be given according to Definition 4.1 and (5.10),(5.11). Then for $t \geq \tau_n > \tau_m$, $\delta < 1/2$

(5.13) $$\|(mx - pt)\ U_D(t,\tau_n)\ U(\tau_n,\tau_m)\ f(p/m)f(x/\tau_m)\| \le$$

$$\leq \text{const}(\tau_m)\cdot t^{\frac{1}{2}-\delta} .$$

Proof.

(5.14)
$$m[x,U'(t,\tau_n)] = U'(t,\tau_n) \int_{\tau_n}^{t} ds\ s\ (\nabla V_\ell)(sp/m)$$

implies with (1.8)

(5.15)
$$\|[x,U'(t,\tau_n)]\ \bar{f}(p/m)\| \leq \text{const.}\ t^{\frac{1}{2}-\delta} .$$

(5.16)
$$\|(mx-pt)\ e^{-iH_0(t-\tau_n)}\ U(\tau_n,\tau_m)\ f\ f\|$$

$$= \|(mx - p\tau_n)\ U(\tau_n,\tau_m)\ f(p/m)\ f(x/\tau_n)\|$$

and (5.3) together imply (5.13). ■

Proposition 5.4. Let $H = H_0 + V_s + V_\ell$ satisfy (1.4) - (1.9). Let f and U be as in Definition 4.1 and U_D satisfy (5.10),(5.11). Then for any m

(5.17)
$$\lim_{\tau_n\to\infty}\ \sup_{T\geq\tau_n}\ \|[e^{-iH(T-\tau_n)}-U_D(T,\tau_n)]\ U(\tau_n,\tau_m)\ f(p/m)f(x/\tau_m)\|$$

$$= 0$$

Proof. By the Cook estimate the sup is bounded by

(5.18)
$$\int_{\tau_n}^{\infty} dt\ \|[V_s+V_\ell(x)-V_\ell(tp/m)]\ U_D(t,\tau_n)\ U(\tau_n,\tau_m)\ f\ f\|$$

$$\leq \int_{\tau_n}^{\infty} dt\{\|V_s(H_0-z)^{-1}\ F(|x| > u_0 t)\|$$

$$+ \|V_s(H_0-z)^{-1}\|\cdot\|F(|x| < u_0 t)\ U_D(t,\tau_n)\ (H_0-z)\ \bar{g}(p/m)\bar{f}(x/\tau_n)\|$$

$$+ \|V_\ell\|\cdot\|F(|x| < u_0 t)\ U_D(t,\tau_n)\ \bar{g}(p/m)\bar{f}(x/\tau_n)\|$$

$$+ \|\{V(t;x)-V(t;tp/m)\}\ U_D(t,\tau_n)\ U(\tau_n,\tau_m)\ f(p/m)f(x/\tau_m)\|\}.$$

Here we have used the shorthand

$$(5.19) \qquad V(t;y) = V_\ell(y)\cdot[1-\varphi(y/u_0t)]$$

with φ as given in (4.3), and the facts that $V_\ell(x) - V(t;x)$ has support in $|x| \le u_0t$ and is norm bounded by $\|V_\ell\|$. Moreover $V_\ell(tp/m) = V(t;tp/m)$ for all values of p in supp $\bar{f}$. The first three summands are integrable in t by (1.7′) and (5.12).

To estimate the difference of the potentials in the last term of (5.18) we write them as Fourier integrals and apply the identity for functions of the operators x and p

$$(5.20) \qquad \exp\{iq(s[x-pt/m]+pt/m)\}$$

$$= \exp\{iq\ pt/m\}\ \exp\{iq\ s[x-pt/m]\}\ \exp\{-its|q|^2/2m\}.$$

Then one obtains

$$(5.21) \qquad V(t;x) - V(t;tp/m)$$

$$= \int_0^1 ds\ \frac{d}{ds}\ V(t;s[x-tp/m] + tp/m)$$

$$= \int_0^1 ds\{(\nabla V)(t;s[x-tp/m] + tp/m)\cdot[x-tp/m]$$

$$+ i(\Delta V)(t;s[x-tp/m] + tp/m)\cdot t/2m\}.$$

The gradient and Laplacian apply to the second argument. Clearly independent of the complicated argument

$$(5.22) \qquad \|(\nabla V)(t;\cdot)\| \le \text{const.}\, t^{-(\delta+3/2)},$$

$$(5.23) \qquad \|(\Delta V)(t;\cdot)\|\cdot t \le \text{const.}\, t^{-(2+\delta)}\cdot t = \text{const.}\, t^{-(1+\delta)}.$$

Thus the contribution with the Laplacian is integrable in t. For the gradient-term we combine (5.22) with (5.13) to conclude integrability. This finishes the proof of Proposition 5.4. ■

The main result of this section is

Theorem 5.5. Let $H = H_0 + V_s + V_\ell$ satisfy (1.4) - (1.9), and let U_D be as defined in (5.10),(5.11). Then for any $\Psi \in \mathcal{H}$

$$(5.24) \qquad \lim_{\tau\to\infty} \sup_{t\geq 0} \|[e^{-iHt} - U_D(\tau+t,\tau)]\, U_D(\tau,0)\Psi\| = 0,$$

and for any $\Psi \in \mathcal{H}^{cont}(H)$

$$(5.25) \qquad \lim_{\tau\to\infty} \sup_{t\geq 0} \|[e^{-iHt} - U_D(\tau+t,\tau)]\, e^{-iH\tau}\Psi\| = 0.$$

Consequently the modified wave operators

$$(5.26) \qquad \Omega_\pm^D = \text{s-}\lim_{t\to\pm\infty} e^{iHt}\, U_D(t,0)$$

exist and are complete, i.e.

$$(5.27) \qquad \text{Ran}\, \Omega_\pm^D = \mathcal{H}^{cont}(H) = \mathcal{H}^{ac}(H).$$

Proof. It is sufficient to verify (5.24) and (5.25) on a total set of vectors for the sequence $\tau_n \to \infty$ as used above. Let the momentum space wave function $\hat{\psi} \in C_0^\infty(\mathbb{R}^\nu)$ satisfy the support property (4.6). Then (5.1) and (5.15) imply

$$(5.28) \qquad \|(mx-pt)\, U_D(t,0)\, \Psi\| \leq \text{const.}\, t^{\frac{1}{2}-\delta}$$

and by Lemma 5.2 for $\beta = 0,1$

$$(5.29) \qquad \|F(|x|<u_0 t)\, U_D(t,0)\, (H_0-z)^\beta\, \Psi\| \leq C_n(1+t)^{-n}.$$

By the proof of Proposition 5.4 (5.24) holds. For (5.25) choose $\Psi = F(E_1 < H < E_2)\Psi$ for some $0 < E_1 < E_2 < \infty$. Choose a finite decomposition $\{f_i\}$ satisfying (2.3) and (2.4) with $i \in \{1,2,...,N\}$ elements. According to Theorem 2.2 there is for any $\varepsilon > 0$ a $T(\varepsilon)$ such that for all $\tau \geqslant T(\varepsilon)$

$$(5.30) \qquad \| e^{-iH\tau}\,\Psi - \sum_{i=1}^{n} f_i(p/m) f_i(x/\tau)\; e^{-iH\tau}\Psi \| < \varepsilon/6.$$

For each f_i following Definition 4.1 construct the intermediate time evolution U (which depends on the finitely many i's). Then by Proposition 4.2 there is a $\tau_m = \tau_m(\epsilon) \geqslant T(\varepsilon)$ such that for all $1 \leq i \leq N$

$$(5.31) \quad \sup_{n>m} \| [e^{-iH(\tau_n-\tau_m)} - U(\tau_n,\tau_m)]\; f_i(p/m) f_i(x/\tau_m) \| < \varepsilon/6N.$$

The estimates (5.30),(5.31) for all $\tau_n > \tau_m$ combine to

$$(5.32) \quad \| e^{-iH\tau_n}\,\Psi - \sum_{i=1}^{N} U(\tau_n,\tau_m) f_i(p/m) f_i(x/\tau_m)\; e^{-iH\tau_m}\,\Psi \| < \varepsilon/3.$$

With Proposition 5.4 we can find $\tau_n > \tau_m$ such that for all i

$$(5.33) \quad \sup_{t\geqslant 0} \| [e^{-iHt} - U_D(\tau_n+t,\tau_n)]\; U(\tau_n,\tau_m)\; f_i(p/m) f_i(x/\tau_m) \| < \varepsilon/3N.$$

Then for this (or larger) τ_n

$$(5.34) \qquad \sup_{t\geqslant 0} \| [e^{-iHt} - U_D(\tau_n+t,\tau_n)]\; e^{-iH\tau_n}\,\Psi \| < \varepsilon.$$

The properties of the modified wave operators are shown as in the short range case of Corollary 3.5. ■

<u>Remarks</u> <u>5.6</u>. (a) If the long-range potential describes the Coulomb force, then it is not necessary to use an intermediate time evolution. One can show directly for suitable Ψ

$$(5.35) \quad \lim_{\tau\to\infty} \sup_{t\geqslant 0} \| [e^{-iHt} - U_D(\tau+t,\tau)]\; f_i(p/m) f_i(x/\tau)\; e^{-iH\tau}\Psi \| = 0.$$

For details see section V of [12] or [37], [49].

(b) The Heisenberg equations of motion have been used already by Alsholm and Kato [1] in their existence proof of modified wave operators. Perry [42,43] used estimates of asymptotic observables to show completeness. Although some technical details are different our proof of Proposition 5.4 is closely related to his. Perry and also Muthuramalingam and Sinha [40], [38] show the bounds on asymptotic observables for the time evolution generated by $H_\ell = H_0 + V_\ell$. See also [26] for related results. Our proof of Proposition 5.1 using the intermediate time evolution U is new. It is simpler and gives stronger results, but most important it carries over easily to higher particle numbers, see Section XII.

VI. More Propagation Properties for the Interacting Time Evolution.

The detailed estimates of the previous sections were obtained for states with energy away from zero such that small speeds were absent. If the energy of a state is bounded above by $m\, v^2/2$ then the asymptotic observables can be used as in Section II to show

$$\lim_{\tau\to\infty} \| F(|x|>v\tau)\, e^{-iH\tau}\, \Psi \| = 0. \tag{6.0}$$

Thus "low energy particles travel slowly". Here we are treating this question in more detail to obtain fast decay in (6.0) for suitable states Ψ. Similar estimates were given for the free time evolution in Proposition 3.1 and as a byproduct of the Definition 4.1 for the time evolution U. Here we show

<u>Theorem 6.1</u>. Let H satisfy (1.4)-(1.9) and let $g \in C_0^\infty(\mathbb{R})$ satisfy

$$\text{supp}\, g \subset (-\infty, (m/2)v^2), \quad v > 0. \tag{6.1}$$

Then uniformly in R

$$\lim_{r\to\infty} \int_0^\infty dt\, \| F(|x|>R+v|t|+r)\, e^{-iHt}\, g(H)\, F(|x|<R) \| = 0. \tag{6.2}$$

If in addition the stronger short-range condition (1.10) holds, then

$$\|F(|x|>R+v|t|+r)\ e^{-iHt}\ g(H)\ F(|x|<R)\| \tag{6.3}$$

$$\leq C_n(1+r+|t|)^{-n} \text{ for all } n \in \mathbb{N}.$$

The constants C_n depend on g and v but are uniform in R.

The theorem will be discussed in detail at the end of this section. As a preparation for the proof we need the technical

Proposition 6.2. Let $H = H_0 + V$ satisfy (1.4)-(1.6) for a multiplication operator $V = V(x)$, and let $g \in C_0^\infty(\mathbb{R})$ or $g(\omega) = (\omega-z)^{-k}$, $z \in \rho(H)$ (the resolvent set), $k \in \mathbb{N}$. Then

a) for any $n \in \mathbb{N}$, uniformly in $\rho \geq 0$

$$\|F(|x|>\rho+r)\ g(H)\ F(|x|<\rho)\| \leq C_n(1+r)^{-n}. \tag{6.4}$$

b) If in addition $\|(H_0+\mathbb{1})^{-1}\ V\ F(|x|>R)\| \leq \text{const}(1+R)^{-\varepsilon}$, $\varepsilon > 0$,

then

$$\|F(|x|>r)\ [g(H)-g(H_0)]^k\| \leq \text{const}.(1+r)^{-\varepsilon k}. \tag{6.5}$$

Remarks 6.3. (i) For the special case $H = H_0$ part a) easily follows using Fourier transformations. Integrable decay was shown in [11] for Hamiltonians with potentials of short and long range. The rapid decay is a result of Krishna [29,30]. (ii) If H satisfies (1.4)-(1.8) then (6.5) holds with $\varepsilon > 1/2$. The decay is faster if better approximations for g(H) are used [11]. (iii) For some weaker sufficient conditions like form sums or non-local potentials see below.

Proof. a) Let ψ_r be any family of functions in $C_0^\infty(\mathbb{R}^\nu)$ depending on ρ and r which satisfies: $0 \leq \psi_r(x) \leq 1$,

$$\psi_r(x) = 1 \text{ if } |x| \le \rho, \tag{6.6}$$

$$\psi_r(x) = 0 \text{ if } |x| \geqslant \rho + r,$$

and there is a constant C such that for any $\rho, r > 0$

$$|\nabla\psi_r(x)| \le c/r. \tag{6.7}$$

From (6.5) it follows that

$$\|F(|x|>\rho+r)\ g(H)\ F(|x|<\rho)\| \tag{6.8}$$

$$= \|-F(|x|>\rho+r)\ [\psi_r(x), g(H)]\ F(|x|<\rho)\|$$

$$= \|F(|x|>\rho+r)\ Ad_r^k\{g(H)\}\ F(|x|<\rho)\| \le \|Ad_r^k\{g(H)\}\|,$$

where we use the shorthand $Ad_r^k\{\cdot\}$ to denote the k-fold commutator with $\psi_r(x)$. To avoid domain questions later it is useful to go from H to its resolvent. For some real

$$a \le \inf \sigma(H) - 1$$

we denote the self-adjoint resolvent by $R = (H-a)^{-1}$, $\|R\| \le 1$. Then we may assume without loss of generality for g(H), $g \in C_0^\infty(\mathbb{R})$, that supp g $\subset (a + \frac{1}{2}, \infty)$.

$$\varphi(\lambda) := g(a+1/\lambda) \tag{6.9}$$

satisfies $\varphi \in C_0^\infty(\mathbb{R}_+)$ and

$$g(H) = \varphi(R). \tag{6.10}$$

We expand φ as its Fourier integral

$$\varphi(R) = (2\pi)^{-\frac{1}{2}} \int dt\ \hat{\varphi}(t)\ e^{itR}. \tag{6.11}$$

$$(6.12) \qquad \mathrm{Ad}_r\{e^{iRt}\} = e^{iR(t-s)}\,\psi_r(x)\,e^{iRs}\Big|_{s=0}^{t}$$

$$= i\int_0^t ds\; e^{i(t-s)R}\,\mathrm{Ad}_r\{R\}\,e^{isR}.$$

By iteration one sees that $\mathrm{Ad}_r^k\{e^{iRt}\}$ is a sum of up to k-fold integrals of products of exponentials multiplied with products of (possibly multiple) commutators $\mathrm{Ad}_r^{\ell}\{R\}$. The total number of Ad_r's adds up to k. The integrations are bounded by $\mathrm{const}(1+|t|)^k$ which can be absorbed by the rapid decay of $\hat{\varphi}(t)$ in (6.11). Therefore it is sufficient to study the behaviour of $\mathrm{Ad}_r^{\ell}\{R\}$ as $r \to \infty$. From here on R may be any resolvent $(H-z)^{-1}$, $z \in \rho(H)$. Then the case where $g(H)$ is a resolvent is covered as well.

$$(6.13) \qquad \mathrm{Ad}_r\{R\} = R\,\mathrm{Ad}_r\{-H\}\,R.$$

By iteration it is clear that the multiple commutator of R is a sum of products of resolvents and (multiple) commutators with H. The commutators of H are always separated by resolvents.

$$(6.14) \qquad R^{1/2}\,\mathrm{Ad}_r\{H\}\,R^{1/2} = R^{1/2}\,\{[\psi_r(x),H_0]+[\psi_r(x),V]\}\,R^{1/2}.$$

All summands extend uniquely to bounded operators (even if the Hamiltonian were defined as a form sum only!). For multiplication operators V the second summand vanishes, the norm of the first is

$$(6.15) \qquad (2m)^{-1}\,\|R^{1/2}\,\{p\cdot(\nabla\psi_r)(x)+(\nabla\psi_r)(x)\cdot p\}\,R^{1/2}\|$$

$$\le \mathrm{const.}\|p\,R^{1/2}\|\,\sup|\nabla\psi_r| \le \mathrm{const}/r.$$

$$(6.16) \qquad \|\mathrm{Ad}_r^2\{H\}\| = \frac{1}{m}\,\||\nabla\psi_r|^2(x)\| \le \mathrm{const}/r^2$$

$$\mathrm{Ad}_r^{\ell}\{H\} = 0 \text{ for } \ell > 2.$$

These estimates and the iteration of (6.13) imply

$$(6.17) \qquad \|\mathrm{Ad}_r^{\ell}\{R\}\| \le \mathrm{const}(\ell)/r^{\ell}.$$

The same estimate holds for powers of resolvents. Thus we have shown

$$(6.18)\qquad \|Ad_r^k\{g(H)\}\| \le \text{const}(k)/r^k.$$

With (6.8) and the uniform boundedness of the norm in (6.4) for small r we have shown part a) of the proposition.

As a particular case of the estimates (6.11)-(6.15) we note

$$(6.19)\qquad \|[g(H),f(x)]\| \le \text{const}(g)\cdot\sup_x|(\nabla f)(x)|.$$

$$(6.20)\ \text{b)}\qquad \|F(|x|>r)\ [g(H)-g(H_0)]^k\|$$

$$\le \prod_{\ell=1}^{k} \|F(|x|>r\ell/k)\ [g(H)-g(H_0)]\|$$

$$+\ (2\|g\|)^{k-1} \sum_{\ell=2}^{k} \|F(|x|>r\ell/k)\ [g(H)-g(H_0)]\ F(|x|<r(\ell-1)/k)\|.$$

Each term in the sum decays rapidly in r by part a). It remains to show that each factor in the product decays as fast as the potentials do. If $g(H) = (H-z)^{-1}$ then

$$(6.21)\qquad \|F(|x|>r)\ [(H-z)^{-1}-(H_0-z)^{-1}]\|$$

$$\le \|F(|x|>r)\ (H_0-z)^{-1}\ F(|x|<r/2)\|\cdot\|V(H-z)^{-1}\|$$

$$+\ \|(H_0-z)^{-1}\|\cdot\|F(|x|>r/2)\ V\ (H_0-z)^{-1}\|\cdot\|(H_0-z)(H-z)^{-1}\|$$

$$\le \text{const}(1+r)^{-\varepsilon}.$$

Similarly for powers of resolvents. For more general functions g see Lemma 2 in [7], Prop. 3.1 in [11], or Lemma 2.3 in [50]. ∎

<u>Remark 6.4</u>. If (part of) the potential V′ is non-local (i.e. not a multiplication operator) it should have rapid decay like typical separable potentials in nuclear physics do. Then the second term in (6.14) is estimated

$$\|R^{1/2}\ [\psi_r(x),V']\ R^{1/2}\| = \|R^{1/2}\ [(1-\psi_r(x)),V']\ R^{1/2}\|$$

$$\leq \|R^{1/2}\ (1-\psi_r(x))\ V'\ R^{1/2}\| + \|R^{1/2}\ V'\ (1-\psi_r(x))\ R^{1/2}\|$$

which has rapid decay in r. Clearly this carries over to higher commutators and the result of Proposition 6.2 is unchanged.

Lemma 6.5. Let the positive uniformly bounded function $h(r,t)$ satisfy the integral inequality for $r \geqslant b|t|$, $|s_k| \leq |t|$, all $\ell \in \mathbb{N}$, some $0 < \alpha < 1$:

$$h((\ell+1)r,t) \leq h_0(r) + h_1(r+|t|) \int_0^t ds\ h(\ell r,s) + \sum_{k=1}^{\infty} h_1(r^{\alpha}+k)\ h(\ell r,s_k), \tag{6.22}$$

where

$$h_0(r) \leq C_n(1+r)^{-n} \text{ for all } n \in \mathbb{N}, \tag{6.23}$$

$$h_1(r) \leq C(1+r)^{-1-\varepsilon},\ \varepsilon > 0. \tag{6.24}$$

Then for $r \geqslant b|t|$

$$h(r,t) \leq C_k(1+r)^{-k} \text{ for all } k \in \mathbb{N}. \tag{6.25}$$

Proof. The result holds if for $r \geqslant b|t|$

$$h(\ell r,t) \leq C(\ell)(\ell r)^{-(\ell-1)\varepsilon\alpha}, \tag{6.26}$$

which we prove by induction for all ℓ. By the uniform boundedness of h it is true for $\ell = 1$. The integral inequality (6.22) implies

$$(\ell r)^{(\ell-1)\varepsilon\alpha}\ h(\ell r,t) \leq (\ell r)^{(\ell-1)\varepsilon\alpha}\ h_0(r) + \tag{6.27}$$

$$+ (\ell r)^{\varepsilon\alpha}\ \{\sum_{k=1}^{\infty} h_1(r^{\alpha}+k)+|t|\ h_1(r+|t|)\}\ (\ell r)^{(\ell-2)\varepsilon\alpha}\ h((\ell-1)r,s)$$

$$\leq C(\ell) < \infty$$

if (6.26) is satisfied for $\ell - 1$. ■

Proof of Theorem 6.1. We show (6.3), for (6.2) see the remarks below. There is a $v' > 0$ with $v' < v$ depending on $\mathcal{g}$ such that the assumption (6.1) of the theorem is satisfied for v'. Then

$$(6.28) \qquad v|t| + r = v'|t| + [(v-v')|t| + r] = v'|t| + r'.$$

If we show for any pair $(\mathcal{g}, v)$ which satisfies (6.1)

$$(6.29) \qquad \|F(|x|>R+v|t|+r)\, e^{-iHt}\, \mathcal{g}(H)\, F(|x|<r)\|$$

$$\leq C_n(b)(1+r)^{-n}, \quad r \geq b|t|, \text{ all } b > 0,\ n \in \mathbb{N}$$

with $C_n(b)$ independent of t, then it holds in particular for v' and $r' = (v-v')|t| + r$. Thus (6.29) implies (6.3). We use the shorthand (omitting the irrelevant dependence on R)

$$(6.30) \qquad h(r,t) = \|F(|x|>r+v|t|+R)\, e^{-iHt}\, \mathcal{g}(H)\, F(|x|<R)\|.$$

By Lemma 6.5 it is sufficient to show for $r \geq b|t|$ uniformly in R the integral inequality

$$(6.31) \qquad h((\ell+1)r,t)$$

$$\leq h_0(r) + \sum_{k=1}^{\infty} h_1(r^\alpha+k)\, h(\ell r, s_k)$$

$$+ h_1(r+t) \int_0^t ds\, h(\ell r, s),$$

where h_0 and h_1 satisfy (6.23) and (6.24), respectively. They may depend on $b > 0$. Then (6.29) and the theorem follow.

Let $g \in C_0^\infty(\mathbb{R})$, $0 \leq g \leq 1$, satisfy

$$(6.32) \qquad g(\omega) = 1 \text{ on } \operatorname{supp} \mathcal{g},$$

$$\mathrm{supp}\; g \subset (-\infty, (m/2)(v-a)^2)$$

for some (small) $a > 0$. Then $g(H)\, \mathring{g}(H) = \mathring{g}(H)$. In addition to the energy cutoff we can introduce a kinetic energy cutoff in our expression.

$$(6.33)\quad \|F(|x|>R+v|t|+(\ell+1)r)\; g(H)\; e^{-iHt}\; \mathring{g}(H)\; F(|x|<R)\|$$

$$\le \sum_{k=1}^{2n-1} \|F(|x|>R+v|t|+(\ell+1)r)\; [g(H)-g(H_0)]^k\; F(|x|<R+v|t|+(\ell+\tfrac{1}{2})r)\|$$

$$+ \|F(|x|>R+v|t|+(\ell+1)r)\; [g(H)-g(H_0)]^{2n}\|$$

$$+ [\sum_{k=1}^{2n-1} 2^k]\; \|F(|x|>R+v|t|+(\ell+\tfrac{1}{2})r)\; g(H_0)\; e^{-iHt}\; \mathring{g}(H)\; F(|x|<R)\|$$

$$\le C_n(1+r)^{-n}$$

$$+ \mathrm{const}(n)\cdot\|F(|x|>R+v|t|+(\ell+\tfrac{1}{2})r)\; g(H_0)\; e^{-iHt}\; \mathring{g}(H)\; F(|x|<R)\|.$$

In the first summand n can be chosen arbitrarily large by Proposition 6.2. Thus we take it as our first contribution to h_0. It remains to show

$$(6.34)\qquad \|F(|x|>R+v|t|+(\ell+\tfrac{1}{2})r)\; g(H_0)\; e^{-iHt}\; \mathring{g}(H)\; F(|x|<R)\|$$

$$\le h_0(r) + \sum_k h_1(r^\alpha+k)\; h(\ell r, s_k)$$

$$+ h_1(r+t) \int_0^t ds\; h(\ell r, s).$$

Our strategy uses the fact that $F(|x| > R + v|t| + (\ell + \frac{1}{2})r)\; g(H_0)$ annihilates all states unless they are far from the origin (where the potential is weak) and have small speeds. According to the intuition from free classical particles for $|s| \le |t|$ the operator

$$F(|x|>R+v|t|+(\ell+\tfrac{1}{2})r)\; g(H_0)\; e^{-iHs}$$

should annihilate all states localized inside a ball of radius $R+v|t|+(\ell+\frac{1}{2})r-(v-2a)|s|$. Taking into account quantum tails one still obtains

$$\|F(|x|>R+v|t|+(\ell+\tfrac{1}{2})r)\ g(H_0)\ e^{-iH_0 s}\ F(|x|<R+v|t|+\ell r-(v-a)|s|)\|$$

$$\leq C_n(1+r+|s|)^{-n} \text{ for all } n.$$

Note that uniformly in R and $|s| \leq |t|$ the motion is restricted to a region which is separated from the origin by $r + a|t|$. Therefore the potentials are weak and their influence can be estimated. If the potential is of short range then the free time evolution is a sufficiently good approximation. The essential estimate in this simpler case is (2.22) of [13]. In the presence of long range potentials, however, a better approximation must be used. It turns out that the "intermediate" approximate time evolution U of Section IV is well suited. We adjust it slightly to the geometry of the present problem.

Again we pick a $1 < \rho < 1/(1-2\delta)$ depending on the decay rate of the long range potential (1.9). The subdivision of the time interval [0,t] is chosen as follows ($t < 0$ is analogous). Let

(6.35) $$m := r^{1/2\rho} \geq 1,$$

$$s_k := (m+k)^{2\rho}-m^{2\rho} = (m+k)^{2\rho}-r,\ k = 0,1,\dots,N-1,$$

$$s_N := t,$$

where N is determined by

(6.36) $$(m+N-1)^{2\rho} - m^{2\rho} < t \leq (m+N)^{2\rho} - m^{2\rho}.$$

$$t_k := s_k - s_{k-1} \leq s_k,\ k = 1,2,\dots,N.$$

They satisfy for $\delta' < \delta$ such that $\rho \leq 1/(1-2\delta')$

(6.37) $$t_k \le \rho(m+k)^{2\rho-1} \le \text{const } (r+s_k)^{\delta'+1/2}, \text{ all } k,$$

(6.38) $$t_k \geqslant \rho(m+k-1)^{2\rho-1} \geqslant \text{const } (m+k)^{2\rho-1}, \quad k = 1,2,\ldots,N-1.$$

A family of spatial cutoff functions f_k will be used at times s_k. Let χ_k be the characteristic function of the exterior of a ball around the origin with radius

$$R + (\ell + \tfrac{1}{4})r + a|t| + (v - a)s_k.$$

We smooth it with ψ as in (4.12)

(6.39) $$f_k := (k+m)^{-\rho} \chi_k * \psi[\cdot/(k+m)^{\rho}].$$

The functions f_k satisfy $0 \le f_k \le 1$, $(1 - f_k) \in C_0^\infty(\mathbb{R}^\nu)$,

(6.40) $$|(\nabla f_k)(x)| \le \text{const.}(k+m)^{-\rho} \text{ for all } x \in \mathbb{R}^\nu,$$

(6.41) $$\operatorname{supp} f_k \subset \{x\in\mathbb{R}^\nu \mid |x|>R+(\ell+\tfrac{1}{4})r + a|t| + (v-a)s_k - (k+m)^{\rho}\},$$

(6.42) $$\operatorname{supp}(1-f_k) \subset \{x\in\mathbb{R}^\nu \mid |x|<R+(\ell+\tfrac{1}{4})r + a|t| + (v-a)s_k + (k+m)^{\rho}\}.$$

In our notation we suppress the dependence of f_k on $R, |t| \geqslant 0$, $\ell \in \mathbb{N}$, because we are mainly interested in estimates uniform in these parameters. By Proposition 3.1 the free time evolution satisfies for all $n \in \mathbb{N}$

(6.43) $$\|f_k(x)\ g(H_0)\ e^{-iH_0(s_k-s)}\ F(|x|<R+\ell r+vs)\|$$

$$\le C_n(1+r+(s_k-s))^{-n}, \quad s_{k-1} \le s \le s_k, \quad k = 1,2,\ldots,N-1,$$

because for $r \geqslant b|t|$, $b > 0$ and $m \geqslant 1$

(6.44) $$\operatorname{dist}(\operatorname{supp} f_k, \{x\in\mathbb{R}^\nu \mid |x|<R+\ell r+vs\}) - (v-2a)(s_k-s) \geqslant$$

$$\geq (r/4) + a|t| + (v-a)s_k - (k+m)^\rho - vs - (v-2a)(s_k-s)$$

$$\geq (r/4) + a(|t|-s_k) + 2a(s_k-s) - (m+N-1)^\rho$$

$$> \mathrm{const}(r+(s_k-s)).$$

In the last step we have used that for $N \geq 2$

$$(r/8) \geq b|t|/8 > (b/8)[(m+N-1)^{2\rho} - m^{2\rho}] \tag{6.45}$$

$$> \mathrm{const}(m+N-1)^{2\rho-1}$$

$$\geq \mathrm{const}(m+N-1)^\rho.$$

Thus the estimate holds uniformly in N (i.e. $|t|$). Similarly one obtains uniformly in $|t|$

$$\|F(|x|>R+(\ell+\tfrac{1}{2})r + v|t|)\, g(H_0)\, e^{-iH_0 t_N}\, [\mathbb{1}-f_{N-1}(x)]\| \tag{6.46}$$

$$\leq C_n(1+r)^{-n}.t$$

Moreover one has for $0 \leq s \leq t$

$$\|F(|x|>R+(\ell+\tfrac{1}{2})r+v|t|)\, g(H_0)\, e^{-iH_0(t-s)}\, F(|x|<R+\ell r+vs+a(t-s))\| \tag{6.47}$$

$$\leq C_n(1+r+(t-s))^{-n},$$

and for $k = 2,3,\ldots,N-1$

$$\|f_k(x)\, g(H_0)\, e^{-iH_0 t_k}\, [\mathbb{1}-f_{k-1}(x)]\| \tag{6.48}$$

$$\leq C_n(m+k)^{-n}.$$

Here we have used the lower bound (6.38) with

$$\text{(6.49)} \qquad \operatorname{dist}\{\operatorname{supp} f_k,\ \operatorname{supp}(1-f_{k-1})\} - (v-2a)t_k$$

$$= at_k - (k+m)^{\rho} - (k-1+m)^{\rho}$$

$$\geq \text{const}(m+k)^{2\rho-1} - 2(k+m)^{\rho}$$

$$\geq \text{const}(m+k)$$

for m large enough depending on ρ only.

Our modified free time evolution takes into account the long range part V_ℓ of the potential. As an approximation of

$$\text{(6.50)} \qquad F(|x|>R+(\ell+\tfrac{1}{2})r+v|t|)\ g(H_0)\ e^{-iH(t-s)}$$

we define for $s_0 = 0 \leq s \leq t = s_N$

$$\text{(6.51)} \qquad U(s) := \begin{cases} F(|x|>R+(\ell+\tfrac{1}{2})r+v|t|)\ g(H_0)\ e^{-iH_0(t-s)}\ e^{-iV_\ell(t-s)} & \text{if } s_{N-1} \leq s \leq s_N \\[2ex] F(|x|>R+(\ell+\tfrac{1}{2})r+v|t|)\ e^{-iH_0 t_N}\ e^{-iV_\ell t_N} \times & \\ \times \prod\limits_{n=k+1}^{N-1}{}' f_n(x)\ e^{-iH_0 t_n}\ e^{-iV_\ell t_n} \times & \\ \times f_k(x)\ g(H_0)\ e^{-iH_0(s_k-s)}\ e^{-iV_\ell(s_k-s)} & \text{if } s_{k-1} \leq s < s_k,\ k = 1,2,\ldots,N-1. \end{cases}$$

Again the product Π' is time ordered, the indices increase from right to left. U(s) is norm differentiable in s for $s_{k-1} < s < s_k$, $k = 1,2,\ldots,N$, and it has discontinuities at s_k, $k = 1,\ldots,N-1$. For $k \leq N-2$

$$\text{(6.52)} \qquad U(s_k) - U(s_k - 0)$$

$$= \ldots f_{k+1}\ g(H_0)\ e^{-iH_0 t_{k+1}}\ e^{-iV_\ell t_{k+1}} -$$

$$- \ldots f_{k+1} e^{-iH_0 t_{k+1}} e^{-iV_\ell t_{k+1}} f_k g(H_0),$$

$$\|U(s_k) - U(s_k - 0)\|$$

$$(6.53) \qquad \le \|f_{k+1} g(H_0) e^{-iH_0 t_{k+1}} [\mathbb{1}-f_k]\|$$

$$+ \|[g(h_0), f_k(x) e^{-iV_\ell t_{k+1}}]\|$$

$$(6.54) \qquad \le C_n(1 + m + k)^{-n} + \text{const}(m + k)^{-\rho}.$$

For the first summand we have used (6.48), for the second (6.40) and (analogous to (4.35))

$$(6.55) \qquad \sup_{x\in \text{supp } f_k} t_{k+1} |\nabla V_\ell(x)| \le (m+k+1)^{2\rho-1}(m+k)^{-\rho(3+2\delta)}.$$

If $k = N - 1$ the first summand in the estimate (6.53) has to be replaced by

$$\|F(|x|>R+(\ell+\tfrac{1}{2})r+v|t|) g(H_0) e^{-iH_0 t_N} [\mathbb{1}-f_{N-1}(x)]\|,$$

which has been estimated in (6.46). Thus the bound (6.54) holds for $k = 1,2,\ldots,N-1$. An additional space cutoff yields the sharper bound uniformly in $|t|$ (or N)

$$(6.56) \qquad \sum_{k=1}^{N-1} \|\{U(s_k) - U(s_k-0)\} F(|x|<R+\ell r+vs_k)\|$$

$$\le C_n(1+r)^{-n}.$$

It follows from (6.43) or (6.47) with (6.45). Now we have collected the technical estimates of the modified free time evolution and we can use it as an approximation of the fully interacting one to obtain (6.34).

$$(6.57) \quad \|F(|x|>R+(\ell+\tfrac{1}{2})r+v|t|) g(H_0) e^{-iHt} \tilde{g}(H) F(|x|<R)\| \le$$

$$\leq \| U(0) \; \mathfrak{g}(H) \; F(|x|<R) \|$$

$$+ \; \| \{ F(|x|>R+(\ell+\tfrac{1}{2})r+v|t|) \; g(H_0) \; e^{-iHt} - U(0) \} \; \mathfrak{g}(H) \; F(|x|<R) \|$$

$$\leq \| U(0) \; F(|x|<R+\ell r) \| + \| F(|x|>R+\ell r) \; \mathfrak{g}(H) \; F(|x|<R) \|$$

$$+ \; \| [U(s) e^{-iHs}]_{s=0}^{t} \; \mathfrak{g}(H) \; F(|x| < R) \| .$$

The first two summands decay rapidly in r by (6.43) and by Proposition 6.2 a), respectively. So they contribute to $h_0(r)$ which has to satisfy (6.23). Since $\exp(-iHs) \; \mathfrak{g}(H)$ is norm differentiable the remaining term is bounded by

$$\sum_{k=1}^{N} \int_{s_{k-1}}^{s_k} ds \; \| \tfrac{d}{ds} \, U(s) \; e^{-iHs} \; \mathfrak{g}(H) \; F(|x|<R) \| \tag{6.58}$$

$$+ \sum_{k=1}^{N-1} \| \{ U(s_k) - U(s_k - 0) \} \; e^{-iHs_k} \; \mathfrak{g}(H) \; F(|x|<R) \| .$$

For $s_{N-1} < s < s_N$ the integrand is bounded by

$$\| F(|x|>R+(\ell+\tfrac{1}{2})r+v|t|) \; g(H_0) \; e^{-iH_0(t-s)} \times \tag{6.59}$$

$$\times \; i[H_0 \; e^{-iV_\ell(t-s)} - e^{-iV_\ell(t-s)} (-V_\ell + H_0 + V_\ell + V_s)] \times$$

$$\times \; e^{-iHs} \; \mathfrak{g}(H) \; F(|x|<R) \|$$

$$\leq \| F(|x|>R+(\ell+\tfrac{1}{2})r+v|t|) \; g(H_0) \; e^{-iH_0(t-s)} \; F(|x|<R+\ell r+vs) \| \times$$

$$\times \; \{ \tfrac{1}{2m} [\| \Delta V_\ell \| (t-s) + (\| \nabla V_\ell \| (t-s))^2 + \| V_s \; \mathfrak{g}(H) \| \}$$

$$+ \sum_j \| F(|x|>R+(\ell+\tfrac{1}{2})r+v|t|) \; g(H_0) \; p_j \; e^{-iH_0(t-s)} \; F(|x|<R+\ell r+ \; vs) \|$$

$$\times \; \tfrac{1}{m} \| \nabla_j V_\ell \| (t - s) \quad +$$

$$+ \|g(H_0)\{[H_0, e^{-iV_\ell(t-s)}] + V_s\}\ F(|x|>R+\ell r+vs)\| \times$$

$$\times \|F(|x|>R+\ell r+vs)\ e^{-iHs}\ \mathring{g}(H)\ F(|x|<R)\|.$$

With the exception of the last product all terms decay faster than any inverse power of τ and $(t - s)$ by (6.47) and their integral contributes to $h_0(r)$. The first factor in the last product of (6.59) is of the type $h_1(r + t)$, i.e. it satisfies (6.24). The cutoff on the right hand side restricts to a region with $|x| > r \geqslant (r + b|t|)/2 \geqslant \mathrm{const}(r + |t|)$. Thus by assumption (1.10)

$$(6.60) \qquad \|g(H_0)\ V_s\ F(|x|>\mathrm{const}(r+t))\| \le \mathrm{const}(1+r+t)^{-1-\varepsilon}.$$

Of the long-range part the gradient-term has the slowest decay

$$(6.61) \qquad (t-s)|(\nabla V_\ell)(x)|\ F(|x|>\mathrm{const}(r+t))$$

$$\le \mathrm{const}\ t_N\ (r+s_N)^{-(\frac{3}{2}+\delta)}$$

$$\le \mathrm{const}\ (r+t)^{-1-\delta+\delta'}$$

by the second inequality in (6.37), $\delta > \delta'$. The square of this expression and the Laplacian-term decay faster. The other integrands satisfy exactly analogous estimates.

$$(6.62) \qquad \|f_k(x)\ g(H_0)\ e^{-iH_0(s_k-s)}\ F(|x|<R+\ell r+vs)\|$$

$$\le C_n(1+r+s_k-s)^{-n}$$

by (6.43) and the same with $g(H_0)p_j$. Moreover

$$(6.63) \quad |x| > R + \ell r + vs \Rightarrow |x| > r \geqslant (r + b|t|)/2 \geqslant \mathrm{const}(r + t).$$

Consequently the sum of the integrals in (6.58) is an integral of the desired form (6.34). For the sum over the discontinuities we use

(6.56) to obtain uniformly in N

(6.64) $$\sum_{k=1}^{N-1} \| \{U(s_k) - U(s_k - 0)\} F(|x|<R+\ell r+vs_k) \|$$

$$\le C_n (1+r)^{-n},$$

i.e. a contribution to $h_0(r)$. Finally with (6.35) $m = r^{\alpha}$, $0 < \alpha = 1/2\rho < 1$ and with $\rho > 1$ the bound (6.54) is of the form $h_1(r^{\alpha} + k)$ and

(6.65) $$\| \{U(s_k)-U(s_k-0)\} F(|x|>R+\ell r+vs_k) e^{-iHs_k} g(H) F(|x|<R) \|$$

$$\le h_1(r^{\alpha}+k) h(\ell r, s_k).$$

Thus with the shorthand (6.30) we have shown the inequality (6.34) and the theorem is proved for $t > 0$. The case $t < 0$ is analogous. ■

Discussion 6.6.

a) The decay rate in (6.0) or (6.2), (6.3) is of limited interest for two-body scattering alone. For our present treatment of three-body systems, however, it is essential to have integrable decay (6.2). The proof is trivial for potentials with $|x|^{-2-\varepsilon}$-decay. For potentials with integrable decay in $|x|$ the result was shown in [11], very severe singularities were admitted. The essentials of the proof for some long range potentials and the basic ideas for the proof given here were mentioned in [11] as well. The proof for potentials with integrable decay was simplified in Section II of [13]. The rapid decay in (6.3) was shown by Krishna [29,30] for potentials with $|x|^{-1-\varepsilon}$-decay, the inclusion of long-range potentials is new here.

b) The proof of integrable decay assuming only (1.7) is a minor variation of the one given here. Since the weaker assumption will hardly matter in applications we omit the details. We expect that the rapid decay (6.3) follows, too, from the integrable decay condition (1.7). For the proof, however, we need the slightly stronger $|x|^{-1-\varepsilon}$-decay (1.10) for the short range part, since Lemma 6.5 does not hold for an h_1 with only integrable decay. For fixed t the rapid decay in r follows from Proposition 6.2 a) since $\exp(-iHt)\mathfrak{g}(H)$ is an admissible function g there. That is true for any local and some non-local potentials (see (6.14) and Remark 6.4.). In Theorem 6.1 the dependence of the constants on t is controlled as well.

c) A straightforward extension to N-body systems has been given by Krishna [29,30]. If subsystems can bind then the relative kinetic energy of subsystems can exceed the total energy by the binding energy. In particular one cannot obtain small values of v by restricting the total energy from above by a small positive bound. Note that the kinematical quantity v came from the support condition on $g(H_0)$ and only indirectly from $\mathfrak{g}(H)$.

d) We will need Theorem 6.1 for small values of v in Sections X and XI below. It is sufficient to have integrable decay, but the rapid decay may be convenient for some estimates.

e) Theorem 6.1 holds on the continuous and point spectral subspaces. We will apply it later both to bound states and scattering states.

f) If the function $\mathfrak{g}$ and the cutoff R are allowed to vary then clearly the vectors in the ranges of $\mathfrak{g}(H)\, F(|x|<R)$ form a dense set of states. Let $\Psi = \mathfrak{g}(H)\, F(|x|<R)\, \Phi$, then by Proposition 6.2 a) with $g(H) = e^{-iHt}\, \mathfrak{g}(H)$ one concludes $e^{-iHt}\, \Psi \in \mathfrak{D}(|x|^k)$ for all k and by Theorem 6.1

(6.66) $$\| |x|^k\, e^{-iHt}\, \Psi \| \le C(n,g,R)\, (1+|t|)^k .$$

This result was obtained by Hunziker [24] for a special class of potentials.

We conclude this section with an amusing application of Proposition 6.2 a). Certainly much better results of this type have been shown by other means.

Corollary 6.7. Let H be such that the result of Proposition 6.2 a) holds (even without uniformity in ρ), and let E be a discrete eigenvalue of H and $H\Psi = E\Psi$. Then

$$\|F(|x|>r)\ \Psi\| \le C_n(\Psi)(1+r)^{-n}.$$

Proof. Let E be non-degenerate. Pick an R such that $(\Psi, F(|x|<R)\Psi) > 0$. There is a $g \in C_0^\infty(\mathbb{R})$ with sufficiently narrow support around E such that

$$g(H)\ F(|x|<R)\ \Psi = \Psi.$$

By (6.4)

$$\|F(|x|>r)\ \Psi\| \le \|\ F(|x|>r)\ g(H)\ F(|x|<R)\| \cdot \|\Psi\|$$

has rapid decay in r. If E is finitely degenerate there is a suitable linear combination of eigenvectors such that for some large enough R

$$\Psi = g(H)\ F(|x|<R)\ \sum_i \lambda_i \Psi_i. \quad \blacksquare$$

Part B. Three-Body Systems.

VII. Introduction and Notation.

We study the motion of three quantum particles in ν-dimensional space. As a general introduction to multiparticle scattering the reader may consult [46]. The particles interact by pair potentials only. Therefore the center of mass motion is free and it can be separated off. We consider here only the more interesting internal motion. There are several different coordinate systems to describe the relative positions of the particles. For our purposes the Jacobi-coordinates are the most convenient ones. We label by α the three possibilities to select a pair out of the three particles and we fix an order for the pair. Then for any pairing α we denote by x^α the relative position of the particles in the pair, the sign is chosen according to the selected order. Then y^α is the position of the third particle relative to the center or mass of the second. If the particles have masses m_i and positions x_i then for the pairing $\alpha = (1,2)$

$$x^\alpha = x_2 - x_1 \tag{7.1}$$

$$y^\alpha = x_3 - (m_1 x_1 + m_2 x_2)/(m_1 + m_2) \tag{7.2}$$

and similarly for the other pairings. A configuration is given uniquely by specifying $\{x^\alpha, y^\alpha\}$ for any α or $\{x^\alpha, x^\beta\}$ or $\{y^\alpha, y^\beta\}$ for some $\alpha \neq \beta$. The different coordinate systems are related by linear transformations, e.g. for $\beta = (3,1)$, $\alpha = (1,2)$

$$x^\beta = -y^\alpha - x^\alpha (m_2/(m_1+m_2)).$$

One always has for $\beta \neq \alpha$

$$x^\beta = \pm y^\alpha + \text{const}\ x^\alpha. \tag{7.3}$$

The state space is the Hilbert space

(7.4) $$\mathcal{H} \cong L^2(\mathbb{R}^{2\nu},\ d^\nu x^\alpha\ d^\nu y^\alpha)$$

of square integrable wave functions $\psi^\alpha(x^\alpha, y^\alpha)$ for some α. In a different coordinate system the wave function transforms to

(7.5) $$\psi^\beta(x^\beta, y^\beta) = \psi^\alpha(x^\alpha(x^\beta, y^\beta), y^\alpha(x^\beta, y^\beta)).$$

We consider ψ^α and ψ^β related by (7.5) as two coordinate system dependent wave functions representing the same state $\Psi \in \mathcal{H}$. The measure in (7.4) is chosen such that the change of coordinates is a unitary map between the L^2-spaces.

The relative momentum operator for the particles in the pair is

(7.6) $$p^\alpha := -i \frac{\partial}{\partial x^\alpha}$$

and for the third particle we have

(7.7) $$q^\alpha := -i \frac{\partial}{\partial y^\alpha}.$$

The reduced masses for the pair and the third particle, respectively, are if $\alpha = (1,2)$

(7.8) $$\mu^\alpha := m_1 \cdot m_2/(m_1 + m_2),$$

(7.9) $$\nu^\alpha := m_3(m_1 + m_2)/(m_1 + m_2 + m_3).$$

Then p^α/μ^α is the relative velocity operator for the particles in the pair and q^α/ν^α is the velocity of the third particle relative to the center of mass of the pair. These kinematical quantities often come up later. The kinetic energy for the internal motion of the pair is

(7.10) $$h_0^\alpha = \frac{1}{2\mu^\alpha}(p^\alpha)^2 = -\frac{1}{2\mu^\alpha}\Delta_{x^\alpha}$$

and for the third particle

$$(7.11)\qquad k_0^\alpha = \frac{1}{2\nu^\alpha}(q^\alpha)^2 = -\frac{1}{2\nu^\alpha}\Delta_{y^\alpha}.$$

The total kinetic energy of the internal motion of the three body system is

$$(7.12)\qquad H_0 = h_0^\alpha + k_0^\alpha,$$

it has the same form for any pairing α! The interaction between the particles is given by three pair potentials V^α. For simplicity we assume that they are multiplication operators in configuration space, i.e. $V^\alpha = V^\alpha(x^\alpha)$. It would be easy to include velocity dependent forces and three-body potentials as well. We will not distinguish in our notation between operators acting on the whole Hilbert space and those acting on a two-body subspace. We assume for all α

$$(7.13)\qquad V^\alpha\ (h_0^\alpha+\mathbb{1})^{-1} \text{ is compact on } L^2(\mathbb{R}^\nu, d^\nu x^\alpha).$$

Then by the Kato-Rellich theorem

$$(7.14)\qquad H = H_0 + V \equiv H_0 + \sum_\alpha V^\alpha$$

is defined as an operator sum and it is self-adjoint on $\mathfrak{D}(H) = \mathfrak{D}(H_0)$. It is easy to include the case where V is Kato-bounded, and with considerable technical efforts many results carry over to form bounded perturbations. We assume that the pair potentials can be decomposed into components of short and of long range

$$(7.15)\qquad V^\alpha = V_s^\alpha + V_\ell^\alpha,$$

$$(7.16)\qquad \|V_s^\alpha\ (h_0^\alpha+\mathbb{1})^{-1}\ F(|x^\alpha|>R)\| \in L^1(\mathbb{R}_+, dR),$$

or equivalently

$$(7.17)\qquad \|(h_0^\alpha+\mathbb{1})^{-1}\ V_s^\alpha\ F(|x^\alpha|>R)\| \in L^1(\mathbb{R}_+, dR).$$

The long-range potential is a differentiable function $V_\ell^\alpha(x^\alpha)$ which tends to zero towards infinity. We assume about the derivative

$$(7.18) \qquad |(\nabla V_\ell^\alpha)(x^\alpha)| \le \mathrm{const}(1+|x^\alpha)^{-(\delta+3/2)}, \quad \delta > 0.$$

In Section XIII we will use this condition with the stronger $\delta > \sqrt{3} - 3/2$. As was discussed in Section I (c.f.(1.9)) we may assume without loss of generality that for all multiindices γ with $|\gamma| \geqslant 1$

$$(7.19) \qquad |(D^\gamma V_\ell^\alpha)(x^\alpha)| \le \mathrm{const}(\gamma)\ (1+|x^\alpha|)^{-1-\delta-|\gamma|/2}.$$

In Sections VIII and IX only the weaker assumptions (8.10) and (8.10′) are needed instead of (7.18) and (7.16), respectively.

$$(7.20) \qquad h^\alpha := h_0^\alpha + V^\alpha$$

is the interacting Hamiltonian for the internal motion of the pair α, and

$$(7.21) \qquad H^\alpha := h^\alpha + k_0^\alpha = H_0 + V^\alpha$$

is the channel Hamiltonian where the interaction between the particles in the pair is kept and the third particle is treated as free.

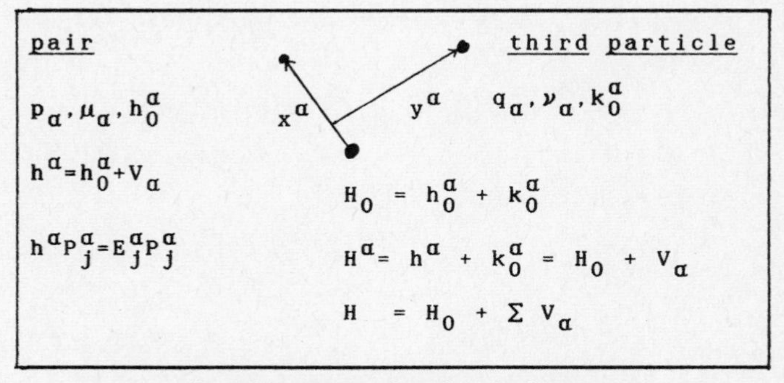

The diagram should help to remember the notation and serve as a quick reference.

For the treatment presented here we have to make an implicit assumption about the bound states of two-body subsystems. We denote by

$$P_j^\alpha$$

the one-dimensional projector in the two body subspace onto the j-th eigenvector (counting multiplicity) of h^α.
Thus

$$(7.22) \qquad h^\alpha P_j^\alpha = E_j^\alpha P_j^\alpha, \quad H^\alpha P_j^\alpha = (E_j^\alpha + k_0^\alpha) P_j^\alpha.$$

The implicit assumption is: for each α, j with $E_j^\alpha = 0$

$$(7.23) \qquad \| F(|x^\alpha| > r) \; P_j^\alpha \| \in L^1(\mathbb{R}_+, dr).$$

If $E_j^\alpha < 0$ the exponential decay of (7.23) is well known (or see Corollary 6.7) and for $E_j^\alpha > 0$ this is the result of Theorem 2.1 in [18]. The assumptions of that theorem are implied by our assumptions or the weaker (8.10), (8.10′). Thus (7.23) holds for all α, j even if we assume it only for $E_j^\alpha = 0$. If only short-range forces are present, then the argument can be modified a bit to avoid this assumption completely. That proof of asymptotic completeness is given in [13]. However, the argument seems to be special for three-body systems (or for particluar N-body systems where any decomposition with a bounded cluster is a two-cluster decomposition.) Therefore we have chosen to present here a treatment which has the property that large portions of it can easily be generalized to higher particle numbers. We believe that the use of (7.23) can be avoided in Sections VIII and IX, but the estimates will be more involved.

The condition (7.13) implies Kato-boundedness and consequently all the operators H_0, H^α, and H are pairwise bounded relative to each other and also

$$h_{(0)}^\alpha \; (H^{(\beta)} - z)^{-1}, \; k_0^\alpha \; (H^{(\beta)} - z)^{-1}$$

are bounded. Analogous to Proposition 6.2 (a) one shows

$$(7.24) \qquad \|F(|x^\beta|<R)\ (h_0^\beta+i)\ (H^{(\alpha)}+i)^{-1}\ F(|x^\beta|>R+r)\| \le C_n/r^n.$$

Indeed, with $\psi_r(x^\beta)$ as given in (6.6), (6.7) the norm is bounded by any multiple commutator

$$(7.25) \qquad \|[\psi_r,\ldots,[\psi_r(x^\beta),(h_0^\beta+i)(H^{(\alpha)}+i)^{-1}]\ldots]\|.$$

$$(7.26) \quad \mu^\beta[\psi_r(x^\beta),(h_0^\beta+i)(H^{(\alpha)}+i)^{-1}]$$

$$= \{1 + (h_0^\beta+i)(H^{(\alpha)}+i)^{-1}\}\ (\nabla\psi_r\cdot p^\beta - \tfrac{i}{2}\ \Delta\psi_r)\ (H^{(\alpha)}+i)^{-1}$$

decays in norm like 1/r and iteration gives (7.24). In particular this can be used to write the short-range condition (7.16) in the following form:

$$(7.27) \qquad \|V_s^\beta\ (H^\alpha+i)^{-1}\ F(|x^\beta|>R)\| \in L^1(\mathbb{R}_+,dR).$$

If a three-body state is orthogonal to a bound state (where all three particles are bounded) then it is suggested by experience from scattering experiments that the state asymptotically breaks up into subsystems which move independently. In contrast to two-body systems, however, there are different possibilities for a simple asymptotic motion, different "scattering channels". In the totally free channel all particles move independently of each other at late times (with proper modifications if long-range forces are present, e.g. between charged particles). In the other channels a pair is in a particular bound state and the third particle moves freely relative to the pair. For these simple systems other possibilities have not been observed in nature. Therefore the mathematical model should have the same properties if suitable conditions are imposed on the forces (i.e. potentials). One says that the three-particle model is "asymptotically complete" if the above list of possible asymptotic behaviour is exhaustive. Then each scattering state can be decomposed into direct summands such that for each component the asymptotic time evolution is simple as given above. Precise definitions are given below in Sections X for the short-range case and XII for long-range potentials. We give a complete proof of asymptotic completeness for potentials which satisfy the above conditions with $\delta > \sqrt{3} - 3/2$, in particular this includes Coulomb forces between charged particles. Most of the results hold if $\delta > 0$ and those of Sections VIII and IX if $\delta > -1/2$. The study of asymptotic observables serves as a tool to control the evolution in phase space of scattering states. In a second step one controls the future time evolution on subsets of phase space which together are absorbing for any scattering state. In the short-range case the asymptotic free time evolution can be used directly. If long-range forces are present, then we use again an intermediate time evolution. It is a good approximation of the true evolution and at the same time it is sufficiently simple such that one can deduce better localization in phase space. That information is sufficient to conclude that the Dollard time evolution is a good asymptotic evolution as long as the energy of the two-body subsystems stays away from zero. Since the intermediate time evolution does not preserve the energy of subsystems one has to give a separate argument to show that energies of subsystems cannot accumulate at zero. That is given in Section XIII.

An outlook as well as references to earlier and related work are given in Section XIV. ■

VIII. Asymptotic Observables for Three-Particle States.

In this section we study the asymptotic behaviour of suitable observables on three-body scattering states. They will be used to control the phase space localization at late times. As in the corresponding Section II where two-body systems were treated the arguments are essentially kinematical. One shows that some potentials do not affect the time evolution of the selected observables. In the totally free channel where all particles separate, all pair interactions can be neglected asymptotically. If a pair is asymptotically bounded then its internal interaction never becomes weak. But the internal motion of a bounded pair is trivial and the net effect on the motion of the third particle is again purely kinematical. The energy which is available for motion is no longer the total energy of the state alone but also the binding energy if bounded pairs are present. Therefore the parts of the state corresponding to different scattering channels evolve under different kinematical conditions. This is the main complication for the three-body system as compared to potential scattering. The fact that infinitely many channels may occur does not cause further problems since zero is the only possible finite accumulation point of eigenvalues for the two body subsystems. Thus within any error margin only finitely many channels have different kinematics.

The main result on asymptotic observables is the following

Theorem 8.1. Let $H = H_0 + \sum_\alpha V^\alpha$ satisfy (7.13)-(7.18), the two-body bound states decay according to (7.23) and let $\Psi \in \mathcal{H}^{cont}(H)$. Then for any $g \in C_0^\infty(\mathbb{R})$, $f \in C_0^\infty(\mathbb{R}^\nu)$, $\tilde{f} \in C_0^\infty(\mathbb{R}^{2\nu})$, and $\varepsilon > 0$ there are an $N = N(\varepsilon)$ and an arbitrary large $\tau = \tau(\varepsilon)$ such that with

(8.1) $$Q^N := \mathbb{1} - \sum_\alpha \sum_{j \le N} P_j^\alpha$$

(8.2) a) $$\| [f(\frac{x^\alpha}{\tau}) - f(\frac{p^\alpha}{\mu^\alpha})] \, Q^N \, e^{-iH\tau} \, \Psi \| < \varepsilon, \text{ all } \alpha;$$

(8.3) b) $$\| f(\frac{y^\alpha}{\tau}) - f(\frac{q^\alpha}{\nu^\alpha}) \, Q^N \, e^{-iH\tau} \, \Psi \| < \varepsilon, \text{ all } \alpha;$$

(8.4) c) $$\| [\tilde{f}(\frac{x^\alpha}{\tau}, \frac{y^\alpha}{\tau}) - \tilde{f}(\frac{p^\alpha}{\mu^\alpha}, \frac{q^\alpha}{\nu^\alpha})] \, Q^N \, e^{-iH\tau} \, \Psi \| < \varepsilon, \text{ all } \alpha;$$

(8.5) d) $$\| [g(h^\alpha) - g(h_0^\alpha)] \, Q^N \, e^{-iH\tau} \, \Psi \| < \varepsilon, \text{ all } \alpha;$$

(8.6) e) $$\| [g(H) - g(H_0)] \, Q^N \, e^{-iH\tau} \, \Psi \| < \varepsilon;$$

(8.7) f) $\sum_{\alpha}\sum_{j\le N} \|[f(\frac{y^\alpha}{\tau}) - f(\frac{q^\alpha}{\nu^\alpha})]\, P_j^\alpha\, e^{-iH\tau}\, \Psi\| < \varepsilon$;

(8.8) g) $\sum_{\alpha}\sum_{j\le N} \{\|[g(H) - g(H^\alpha)]\, P_j^\alpha\, e^{-iH\tau}\, \Psi\|$

$+ \|P_j^\alpha\, [g(H) - g(H^\alpha)]\, e^{-iH\tau}\, \Psi\|\} < \varepsilon$.

For any sequence ε_n the sequence of times $\tau_n = \tau(\varepsilon_n)$ can be chosen such that.

$$\text{(8.9)} \qquad \underset{n\to\infty}{\text{w-lim}}\; e^{-iH\tau_n}\, \Psi = 0.$$

<u>Remarks</u> <u>8.2</u>. a) Actually the long-range potential may have slower decay. In this and the next section we will only use the condition

$$\text{(8.10)} \qquad \lim_{|x|\to\infty} |x|\cdot|(\nabla V_\ell^\alpha)(x)| = 0$$

which is weaker than (7.18). Similarly for the short-range potential compactness of

$$\text{(8.10')} \qquad (1+|x^\alpha|)\; V_s^\alpha\; (h_0^\alpha+\mathbb{1})^{-1}$$

in the two-body subspace is sufficient here. Locally the assumptions may be weakened further.

b) At a late time τ_n we decompose the state $\exp(-iH\tau_n)\,\Psi$ according to

$$\mathbb{1} = Q^N + \sum_{\alpha}\sum_{j\le N} P_j^\alpha$$

into the parts with bounded pairs and a remainder where all pairs are unbounded or only weakly bounded. Since $P_i^\alpha P_j^\beta$ is compact if $i \ne j$ or $\alpha \ne \beta$ the weak convergence to zero (8.9) implies that the decomposition is asymptotically orthogonal. On the parts of the state with a bounded pair the position of the third particle is well correlated to its momentum. The third particle moves relative to the center of mass of the bounded pair exactly as a two body system would behave ("two cluster motion $\hat{=}$ two body motion"). If no pair is bounded then all relative positions and relative momenta are correlated in such a way as if all particles would start at time zero from the origin and travel with constant velocities. Since we do not study $x^{\alpha} - tp^{\alpha}/\mu^{\alpha}$ but instead the more modest question $x^{\alpha}/t - p^{\alpha}/\mu^{\alpha}$ etc., the uncertainty of the initial state and the bending of the trajectories due to interactions both drop out asymptotically. Finally, if the pair α is in the k-th bound state, $k > N$ large, then the internal wave function for the pair is mainly localized where $|x^{\alpha}|$ is large and $h^{\alpha} \approx h_0^{\alpha}$. Therefore the small binding energy implies small p^{α} for these states. Thus within a small finite error $p^{\alpha}/\mu^{\alpha} \approx x^{\alpha}/t$ because the latter tends to zero asymptotically. This explains why it is not necessary to treat all the infinitely many channels separately.

c) Related results have been obtained before [13],[14]. We have eliminated here several technical assumptions used there. The main new result is that all channels can be treated simultaneously and that a common time τ can be found such that all of the above statements hold. This eliminates the need to use absolute continuity as was done in Section III of [14]. Moreover, in our present proof we do not use any information about the further asymptotic time evolution in any of the channels. This makes it easy to generalize the argument to higher particle numbers.

We will give several Lemmas which combine to a proof of Theorem 8.1. Unless stated otherwise we always assume that the assumptions of that theorem (or the weaker (8.10)) are satisfied.

Lemma 8.3. Let H be any self-adjoint operator and C be any compact operator. Then there is $T(\varepsilon)$ such that

$$(8.11) \qquad \sup_{\tau \in \mathbb{R}} \frac{1}{T} \int_{\tau}^{\tau+T} dt \, \|C \, e^{-iHt} \, \Psi\| \le \|C\| \, \|P^{pp}(H)\Psi\| + \varepsilon$$

for all $T \geq T(\varepsilon)$, $\|\Psi\| = 1$. $T(\varepsilon)$ depends on C and H but is independent of Ψ.

Proof. By the Schwarz inequality it is sufficient to estimate uniformly in τ and Ψ

$$(8.12) \quad \frac{1}{T}\int_\tau^{\tau+T} dt\ \|C\ e^{-iHt}\ P^{cont}(H)\ \Psi\|^2$$

$$= (P^{cont}(H)\ e^{-iH\tau}\Psi,\ \frac{1}{T}\int_0^T dt\ e^{iHt}\ C^*C\ e^{-iHt}\ P^{cont}(H)\ e^{-iH\tau}\Psi)$$

$$\leq \|\frac{1}{T}\int_0^T dt\ e^{iHt}\ C^*C\ e^{-iHt}\ P^{cont}(H)\|$$

By (2.27) this vanishes as $T \to \infty$ because C^*C is compact. ■

As an application of this result one can show

Proposition 8.4. Let $H = H_0 + \sum_\alpha V^\alpha$ satisfy (7.13). Then for any r, ε there is a $T(r,\varepsilon)$ such that for all $\Psi \in \mathfrak{D}(H)$, b = 0 or 1, $T \geq T(r,\varepsilon)$

$$(8.13) \quad \sup_{\tau\in\mathbb{R}} \frac{1}{T}\int_\tau^{\tau+T} dt\ \|F(|x^\alpha|<r)\ (h^\alpha)^b\ P^{cont}(h^\alpha)\ e^{-iHt}\Psi\|$$

$$< \varepsilon\|(H+i)\ \Psi\| + \|(H+i)\ P^{pp}(H)\ \Psi\|$$

For b = 0 this is the content of Proposition 3.5 in [13] with $\exp(-iH\tau)\Psi$ inserted for Ψ. The proof is given there for bounded potentials. The generalization to potentials which satisfy (7.13) as well as the inclusion of the case b = 1 are trivial.

Lemma 8.5. Let H be as in Proposition 8.4 and let $\Psi \in \mathcal{H}^{cont}(H)$, $\|\Psi\| = 1$. For any $g \in C_0^\infty(\mathbb{R})$ and $\varepsilon > 0$ there is an $N(g,\varepsilon)$ and a $T(g,N,\varepsilon)$ such that for all $N \geq N(g,\varepsilon)$, $T \geq T(g,N,\varepsilon)$, all α,

$$(8.14)\ a) \quad \sup_{\tau\in\mathbb{R}} \frac{1}{T}\int_\tau^{\tau+T} dt\ \|[g(h^\alpha)-g(h_0^\alpha)]\ Q^N\ e^{-iHt}\Psi\| < \varepsilon,$$

$$(8.15)\ \text{b)} \qquad \sup_{\tau\in\mathbb{R}} \frac{1}{T}\int_{\tau}^{\tau+T} dt\ \|[g(H)-g(H_0)]\ Q^N\ e^{-iHt}\Psi\| < \varepsilon .$$

Proof. We decompose the expressions into a part which is small uniformly in t and a remainder which decays in the time average. It is convenient to rewrite Q^N as

$$(8.16) \qquad Q^N = P^{cont}(h^\alpha) + \sum_{j>N} P_j^\alpha - \sum_{\beta\neq\alpha}\sum_{j\leq N} P_j^\beta .$$

Depending on g and ε determine r such that

$$(8.17) \qquad \|[g(h^\alpha)-g(h_0^\alpha)]\ F(|x^\alpha|>r)\| < \epsilon/12$$

$$(8.18) \qquad \|[g(H)-g(H_0)]\ \prod_\alpha F(|x^\alpha|>r)\| < \varepsilon/3 .$$

Then choose $N(g,\epsilon)$ such that for all $N \geqslant N(g,\varepsilon)$

$$(8.19) \qquad 2\|g\|\cdot \sum_\alpha \|F(|x^\alpha|<r) \sum_{j>N} P_j^\alpha\| < \varepsilon/3 .$$

Thus with $\|Q^N\| \leq 4$ for all N

$$(8.20) \qquad \|[g(h^\alpha)-g(h_0^\alpha)]\ Q^N\ e^{-iHt}\ \Psi\|$$

$$\leq 4\|[g(h^\alpha)-g(h_0^\alpha)]\ F(|x^\alpha|>r)\|$$

$$+\ 2\|g\|\ \|F(|x^\alpha|<r) \sum_{j>N} P_j^\alpha\|$$

$$+\ 2\|g\|\{\|F(|x^\alpha|<r)\ P^{cont}(h^\alpha)\ e^{-iHt}\Psi\|$$

$$+\ \|F(|x^\alpha|<r) \sum_{\beta\neq\alpha}\sum_{j\leq N} P_j^\beta\ e^{-iHt}\ \Psi\|\} .$$

The first two terms are bounded by $\varepsilon/3$ independent of t, the third decays in the time average by Proposition 8.5 and the last by Lemma 8.4 since

$$(8.21) \qquad F(|x^\alpha|<r) \sum_{\beta\neq\alpha} \sum_{j\leq N} P_j^\beta$$

is compact for any N. This proves (a). For (b) we use

$$(8.22) \qquad \|[g(H)-g(H_0)]\, Q^N\, e^{-iHt}\, \Psi\|$$

$$\leq 4\|[g(H)-g(H_0)] \prod_\alpha F(|x^\alpha|>r)\|$$

$$+ 2\|g\| \sum_\alpha \|F(|x^\alpha|<r)\, Q^N\, e^{-iHt}\, \Psi\|$$

and proceed in the same way. ■

Together with H_0 (7.12) the following operators are useful for kinematical considerations

$$(8.23) \qquad D := \tfrac{1}{2}(p^\alpha\cdot x^\alpha+x^\alpha\cdot p^\alpha) + \tfrac{1}{2}(q^\alpha\cdot y^\alpha+y^\alpha\cdot q^\alpha)$$

$$=: D^\alpha + d^\alpha$$

$$= p^\alpha\cdot x^\alpha+q^\alpha\cdot y^\alpha+i\nu$$

$$= x^\alpha\cdot p^\alpha+y^\alpha\cdot q^\alpha-i\nu$$

$$(8.24) \qquad X^2 := \mu^\alpha|x^\alpha|^2 + \nu^\alpha|y^\alpha|^2.$$

Like H_0 all operators are independent of the chosen pairing (coordinate system) α. It is useful that they can be split for any α into a sum of terms which depend only on the internal motion of the pair or on the motion of the third particle alone.

Lemma 8.6. Let H be as in Theorem 8.1.

a) The dense set $\mathfrak{D}' = \mathfrak{D}(H_0) \cap \mathfrak{D}(X^2) \subset \mathfrak{D}(D)$ satisfies

$$(8.25)\quad e^{-iHs}\ \mathfrak{D}' = \mathfrak{D}';$$

$$(8.26)\quad \|(1+X^2)^{\mu/2}\ e^{-iHs}\ \Psi\| \le \mathrm{const}(\Psi)\cdot(1+|s|)^{\mu},\ \mu = 1,2,\ \Psi \in \mathfrak{D}';$$

$$(8.27)\quad (H-z)^{-1}\ \mathfrak{D}' \subset \mathfrak{D}' \text{ for all } z \in \mathbb{C},\ \mathrm{Im}\ z \ne 0.$$

b) The quadratic form

$$(8.28)\qquad K = i[H,D] - 2H_0$$

which is termwise defined on $\mathfrak{D}' \times \mathfrak{D}'$ has the property that for any $z \in \mathbb{C}$, $\mathrm{Im}\ z \ne 0$

$$(8.29)\qquad C = (H-z)^{-1}\ K\ (H-z)^{-1} = \sum_{\alpha} C^{\alpha}$$

extends to a bounded operator and for each α

$$(8.30)\qquad \lim_{r\to\infty}\|C^{\alpha}\ F(|x^{\alpha}|>r)\| = 0.$$

Proof. Part (a) is the exact analogue of the two-body result in Lemma 2.3 and we omit the proof.

b) As a quadratic form on $\mathfrak{D}' \times \mathfrak{D}'$

$$(8.31)\qquad K = i\ [H_0,D] + \sum_{\alpha}\{i[V_{\ell}^{\alpha},D] + iV_s^{\alpha}\ D - iD\ V_s^{\alpha}\} - 2H_0$$

$$= \sum_{\alpha}\{i[V_{\ell}^{\alpha},D^{\alpha}] + iV_s^{\alpha}\ D^{\alpha} - iD^{\alpha}\ V_s^{\alpha}\}$$

$$= \sum_{\alpha}\{-x^{\alpha}\cdot\nabla V_{\ell}^{\alpha} + iV_s^{\alpha}\ x^{\alpha}\cdot p^{\alpha} - ip^{\alpha}\cdot x^{\alpha}\ V_s^{\alpha} + \nu\ V_s^{\alpha}\},$$

where we have used that $i[H_0,D] = 2H_0$. For each α

$$(8.32) \quad \tilde{C}^\alpha = (h_0^\alpha - z)^{-1}\{-x^\alpha \cdot \nabla V_\ell^\alpha + iV_s^\alpha \; x^\alpha \cdot p^\alpha - ip^\alpha \cdot x^\alpha \; V_s^\alpha + \nu V_s^\alpha](h_0^\alpha - z)^{-1}$$

is compact in the two-body factor space as in Lemma 2.3 and for $\tilde{C}^\alpha$ the decay property (8.30) follows. Since for any R by (7.24)

$$(8.33) \quad \lim_{r\to\infty} \|F(|x^\alpha|<R) \; (h_0^\alpha - z)(H-z)^{-1} \; F(|x^\alpha|>R+r)\| = 0$$

the decay (8.30) follows for

$$(8.34) \quad C^\alpha = (H-z)^{-1} \; (h_0^\alpha - z) \; \tilde{C}^\alpha \; (h_0^\alpha - z) \; (H-z)^{-1}. \quad \blacksquare$$

Clearly it is sufficient to verify Theorem 8.1 for a set of vectors which is dense in $\mathcal{H}^{cont}(H)$. Typically the time invariant set

$$(8.35) \quad \mathfrak{D} := (\bigcup_E F(H<E) \; \mathcal{H}^{cont}(H)) \cap (H-z)^{-1} \; \mathfrak{D}'$$

will be dense in $\mathcal{H}^{cont}(H)$ and we will consider below only vectors $\Psi \in \mathfrak{D}$. In the general case when this set might possibly not be dense one can apply exactly the same approximations as in Section II to obtain the same results. Therefore we will not repeat this technicality except for a few remarks below about additional small terms which would occur.

Lemma 8.7. Let the assumptions of Theorem 8.1 be satisfied. Suppose that for any $\Psi \in \mathfrak{D}$ (as given in (8.35)) and $\varepsilon > 0$ there is an $N'(\varepsilon)$ and for each $T > 0$, $N \geq N'(\varepsilon)$ there is an arbitrarily large $\tau = \tau(\varepsilon, T, N)$ with

$$(8.35') \quad \frac{1}{T}\int_\tau^{\tau+T} dt \; \{\frac{\mu^\alpha}{2}\|(\frac{x^\alpha}{t} - \frac{p^\alpha}{\mu^\alpha}) \; Q^N \; e^{-iHt} \; \Psi\|^2$$

$$+ \frac{\nu^\alpha}{2}\| \; (\frac{y^\alpha}{t} - \frac{q^\alpha}{\nu^\alpha}) \; Q^N \; e^{-iHt} \; \Psi\|^2$$

$$+ \sum_\alpha \sum_{j\leq N} \frac{\nu^\alpha}{2}\|(\frac{y^\alpha}{t} - \frac{q^\alpha}{\nu^\alpha}) \; P_j^\alpha \; e^{-iHt} \; \Psi\|^2\} < \varepsilon$$

then Theorem 8.1 holds.

Proof. By Lemma 2.6 there is for given $\varepsilon, f, \tilde{f}$ an $\varepsilon' > 0$ and $\bar{\tau}$ such that

$$(8.36) \quad \frac{\mu^\alpha}{2}\|(\frac{x^\alpha}{\tau} - \frac{p^\alpha}{\mu^\alpha})\, Q^N\, e^{-iH\tau}\Psi\|^2$$

$$+ \frac{\nu^\alpha}{2}\|(\frac{y^\alpha}{\tau} - \frac{q^\alpha}{\nu^\alpha})\, Q^N\, e^{-iH\tau}\,\Psi\|^2$$

$$+ \sum_\alpha \sum_{j\le N} \frac{\nu^\alpha}{2}\|(\frac{y^\alpha}{\tau} - \frac{q^\alpha}{\nu^\alpha})\, P_j^\alpha\, e^{-iH\tau}\,\Psi\|^2 < \varepsilon'$$

for some $t \geqslant \bar{\tau}$ implies that (8.2)-(8.4) and (8.7) are satisfied for this t. If for some τ,T the time average from τ to τ + T of the sum of positive terms (8.5), (8.6), (8.8), and (8.36) is bounded by $\varepsilon' \le \varepsilon$ then there is a time $t \in [\tau, \tau+T]$ such that all terms (8.5), (8.6), (8.8), and (8.36) are bounded by ε' and consequently all bounds (8.2)-(8.8) are satisfied for that t.

Let C_k be a sequence of compact operators with

$$(8.37) \quad \underset{k\to\infty}{\text{s-lim}}\, C_k = \mathbb{1}.$$

If for given ε_n and corresponding ε'_n we include in our time average the fifth positive term

$$(8.38) \quad \|C_n\, e^{-iHt}\Psi\|$$

and determine the appropriate t_n then also

$$(8.39) \quad \|C_n\, e^{-iHt_n}\,\Psi\| < \varepsilon'_n \to 0$$

and the weak convergence to zero (8.9) follows as well.

With $N(g,\varepsilon)$ as given in Lemma 8.5 and $N'(\varepsilon)$ from the assumption of this Lemma set

$$N = N(\varepsilon) := \max\{N(g,\varepsilon'/5), N'(\varepsilon'/5)\} \tag{8.40}$$

and keep it fixed in the sequel. Then for all sufficiently large T by Lemma 8.5

$$\sup_{\tau} \frac{1}{T} \int_{\tau}^{\tau+T} dt \ \|[g(h^{\alpha})-g(h_0^{\alpha})] \ Q^N \ e^{-iHt}\Psi\| < \varepsilon'/5, \tag{8.41}$$

$$\sup_{\tau} \frac{1}{T} \int_{\tau}^{\tau+T} dt \ \|[g(H)-g(H_0)] \ Q^N \ e^{-iHt}\Psi\| < \varepsilon'/5. \tag{8.42}$$

Since $[g(H) - g(H^{\alpha})]P_j^{\alpha}$ and $P_j^{\alpha}[g(H) - g(H^{\alpha})]$ are compact for all α, j we conclude with Lemma 8.3 that for sufficiently large T

$$\sup_{\tau} \frac{1}{T} \int_{\tau}^{\tau+T} dt \sum_{\alpha} \sum_{j\leq N} \{\|[g(H)-g(H^{\alpha}] \ P_j^{\alpha} \ e^{-iHt}\Psi\| + \tag{8.43}$$

$$+ \|P_j^{\alpha} \ [g(H)-g(H^{\alpha})] \ e^{-iHt}\Psi\|\} < \varepsilon'/5,$$

and for an arbitrarily chosen C_k from the sequence (8.37)

$$\sup_{\tau} \frac{1}{T} \int_{\tau}^{\tau+T} dt \ \|C_k \ e^{-iHt}\Psi\| < \varepsilon'/5. \tag{8.44}$$

Now choose a T such that (8.41)-(8.44) hold simultaneously and keep it fixed in the sequel. Finally we choose a large $\tau(\varepsilon'/5,T,N) > \bar{\tau}$ according to the assumption to satisfy (8.35) with $\varepsilon'/5$. By the discussion following (8.36) this implies the statement of Theorem 8.1. ■

Clearly it would have been sufficient to find for any $\Psi \in \mathcal{H}^{cont}(H)$ a Ψ' with $2\|g\| \ \|\Psi-\Psi'\| < \varepsilon$ and Ψ' satisfies (8.35). This can be used if $\mathfrak{D}$ is not dense in $\mathcal{H}^{cont}(H)$. For the proof of Theorem 8.1 it remains to verify the assumptions of Lemma 8.7.

The decay property (7.23) of the eigenfunctions of pairs implies that for any N

(8.45) $$Q^N X^2 Q^N \text{ and } Q^N D Q^N$$

are defined as quadratic forms on $\mathfrak{D} \times \mathfrak{D}$. Similarly the commutators of H with the two products are defined termwise. We calculate them further and obtain:

Lemma 8.8. Let H be as in Theorem 8.1 and $\Psi \in \mathfrak{D}$.

(8.46) a) $$(1+|x^\alpha|)\, i[H,Q^N]\, (H-z)^{-1} \text{ and}$$

(8.47) $$|y^\alpha|\, i[H,Q^N]\, (H-z)^{-1}$$

are compact for all α, N.

b) For any N

(8.48) $$\lim_{\tau\to\infty} \{(\Psi,\, e^{iH\tau}\, Q^N \tfrac{1}{2} \tfrac{X^2}{\tau^2}\, Q^N\, e^{-iH\tau}\Psi)$$

$$- \frac{1}{\tau^2} \int_0^\tau dt\, (\Psi,\, e^{iHt}\, Q^N D\, Q^N\, e^{-iHt}\Psi)\} = 0.$$

Proof. Since Q^N maps $\mathfrak{D}(H)$ into itself the commutators are defined termwise and one obtains

(8.49) $$i[H,Q^N]\, (H-z)^{-1} = \sum_\alpha \sum_{\beta\neq\alpha} i[V^\beta, P_N^\alpha]\, (H-z)^{-1} =: \sum_\alpha K^\alpha$$

where we use the shorthand

(8.50) $$P_N^\alpha := \sum_{j\leq N} P_j^\alpha.$$

Decomposing the potential into its parts of long and short range gives for (8.49)

$$(8.51)\quad K^{\alpha} = \sum_{\beta\neq\alpha} \{iV_s^{\beta}\, P_N^{\alpha} - iP_N^{\alpha}\, V_s^{\beta} + i[(V_\ell^{\beta}(x^{\beta})-V_\ell^{\beta}(\pm y^{\alpha})),P_N^{\alpha}]\}\ (H-z)^{-1}.$$

In the last term we have used that $V_\ell^{\beta}(\pm y^{\alpha})$ commutes with P_N^{α}, the sign is chosen according to (7.3). By the decay of eigenfunctions (7.23) it is easy to see that

$$(8.52)\quad \lim_{R\to\infty} \|F(|x^{\alpha}|+|y^{\alpha}|>R)\ (1+|x^{\alpha}|)\ P_N^{\alpha}\ V_s^{\beta}\ (H-z)^{-1}\| = 0,$$

$$(8.53)\quad \lim_{R\to\infty} \|F(|x^{\alpha}|+|y^{\alpha}|>R)\ P_N^{\alpha}\ |y^{\alpha}|\ V_s^{\beta}\ (H-z)^{-1}\|$$

$$\leq \lim_{R\to\infty} \|F(|x^{\alpha}|+|y^{\alpha}|>R)\ P_N^{\alpha}\ (|x^{\beta}|V_s^{\beta})\ (H-z)^{-1}\|$$

$$+ \lim_{R\to\infty} \|F(|x^{\alpha}|+|y^{\alpha}|>R)\ P_N^{\alpha}\ |\lambda x^{\alpha}|\ V_s^{\beta}\ (H-z)^{-1}\| = 0,$$

where $x^{\beta} = \pm y^{\alpha} + \lambda x^{\alpha}$. With smooth cutoff functions and a bit more effort for commuting the same follows for the term with $V_s^{\beta}\, P_N^{\alpha}$. For the long-range part in (8.51) we do not use the commutator. $(V_\ell^{\beta}(\pm y^{\alpha}+\lambda x^{\alpha}) - V_\ell^{\beta}(\pm y^{\alpha}))$ is uniformly bounded and it decays for large $|y^{\alpha}|$ uniformly for x^{α} in bounded sets. Thus

$$(8.54)\quad \lim_{R\to\infty}\|F(|x^{\alpha}|+|y^{\alpha}|>R)\ (V_\ell^{\beta}(x^{\beta})-V_\ell^{\beta}(\pm y^{\alpha}))\ (1+|x^{\alpha}|)\ P_N^{\alpha}\| = 0.$$

By the decay of the derivative of V_ℓ^{β} (7.18) or (8.10)

$$(8.55)\quad \frac{|y^{\alpha}|}{1+|x^{\alpha}|}[V_\ell^{\beta}(\pm y^{\alpha}+\lambda x^{\alpha}) - V_\ell^{\beta}(\pm y^{\alpha})]$$

is uniformly bounded and it decays in $|y^{\alpha}|$ uniformly for x^{α} in a bounded set. Thus also

$$(8.56)\quad \lim_{R\to\infty}\|F(|x^{\alpha}|+|y^{\alpha}|>R)\ |y^{\alpha}|\ (V_\ell^{\beta}(x^{\beta})-V_\ell^{\beta}(\pm y^{\alpha}))\ P_N^{\alpha}\| = 0.$$

Since x^γ, y^γ are linear combinations of x^α, y^α and

$$F(|x^\alpha|+|y^\alpha|<R)\ K^\alpha \tag{8.57}$$

is compact for any $R < \infty$ we have shown

$$K^\alpha,\ |x^\alpha|K^\alpha,\ \text{and}\ |y^\gamma|K^\alpha \tag{8.58}$$

are compact for all α, γ. This proves (a).

(b)

$$(\Psi, e^{iH\tau}\ Q^N \tfrac{1}{2} \frac{X^2}{\tau^2} Q^N\ e^{-iH\tau}\Psi) \tag{8.59}$$

$$= \frac{1}{\tau^2}(\Psi, Q^N\ X^2\ Q^N\Psi)$$

$$+ \frac{1}{\tau^2} \int_0^\tau dt(\Psi, e^{iHt}\ i[H,\ Q^N \tfrac{1}{2} X^2\ Q^N]\ e^{-iHt}\Psi).$$

Clearly the first term vanishes as $\tau \to \infty$. We expand the commutator and obtain

$$(\Psi,\ e^{iHt}\ Q^N\ i[H, \tfrac{1}{2} X^2]\ Q^N\ e^{-iHt}\Psi) \tag{8.60}$$

$$+ (\Psi,\ e^{iHt}\ Q^N \tfrac{1}{2} X^2\ i[H,Q^N]\ (H-z)^{-1}\ e^{-iHt}(H-z)\Psi)$$

$$+ ((H-z)\Psi,\ e^{iHt}(H-z)^{-1}\ i[H,Q^N] \tfrac{1}{2} X^2\ Q^N\ e^{-iHt}\Psi).$$

The first term in (8.60) is the desired term by $i[H,X^2] = 2D$. The time average of the second is bounded by

$$\sup_{0\le t\le\tau} \frac{1}{2\tau} \| |X|\ Q^N\ e^{-iHt}\Psi\| \cdot \frac{1}{\tau} \int_0^\tau dt\ \|C\ e^{-iHt}(H-z)\Psi\|. \tag{8.61}$$

The first factor is uniformly bounded by Lemma 8.6 (a). $C = |x|i[H,Q^N](H-z)^{-1}$ is compact by part (a) of this Lemma and consequently the time average vanishes as $\tau \to \infty$ since $\Psi \in \mathcal{H}^{cont}(H)$. Analogously for the third summand in (8.45). ∎

Without the simplifying density assumption (8.35) one would have an additional arbitrarily small term

(8.62) $$\mathrm{const}\|C\ P^{pp}(H)(H-z)\Psi'\|$$

and similar corrections in the next Lemmas. They do not effect the final result.

<u>Lemma 8.9</u>. For every $\varepsilon > 0$ and any $\Psi \in \mathfrak{D}$ there is an $N'(\varepsilon) < \infty$ such that for all $N \geq N'(\varepsilon)$

(8.63) $$\limsup_{\tau\to\infty} \{(\Psi, e^{iH\tau}\ Q^N\ \frac{D}{\tau}\ Q^N\ e^{-iH\tau}\Psi)$$

$$-\frac{1}{\tau}\int_0^\tau dt\ (\Psi, e^{iHt}\ Q^N\ 2H_0\ Q^N\ e^{-iHt}\Psi)\} < \varepsilon/4.$$

<u>Proof</u>.

(8.64) $$(\Psi, e^{iH\tau}\ Q^N\ \frac{D}{\tau}\ Q^N\ e^{-iH\tau}\Psi)$$

$$= \frac{1}{\tau}(\Psi, Q^N\ D\ Q^N\ \Psi)$$

$$+ \frac{1}{\tau}\int_0^\tau dt\ (\Psi, e^{iHt}\ i[H, Q^N D Q^N]\ e^{-iHt}\Psi).$$

The first summand vanishes as $\tau \to \infty$, the integrand in the second equals

(8.65) $$(\Psi, e^{iHt}\ Q^N\ i[H,D]\ Q^N\ e^{-iHt}\Psi)$$

$$+ (\Psi, e^{iHt} i[H,Q^N]\ D\ Q^N\ e^{-iHt}\Psi)$$

$$+ (\Psi, e^{iHt}\ Q^N\ D\ i[H,Q^N]\ e^{-iHt}\Psi).$$

The last term is bounded by (see (8.23))

(8.66) $$\||p^\alpha|\ Q^N\ e^{-iHt}\Psi\|\cdot\||x^\alpha| i[H,Q^N]\ (H-z)^{-1}\ e^{-iHt}\ (H-z)\Psi\| +$$

$$+\| |q^\alpha| \; Q^N \; e^{-iHt}\Psi\| \cdot \| |y^\alpha| \; i[H,Q^N] \; (H-z)^{-1} \; e^{-iHt}(H-z)\Psi\|$$

$$+ \; \nu\|Q^N \; e^{-iHt}\Psi\| \cdot \|i[H,Q^N] \; (H-z)^{-1} \; e^{-iHt}(H-z)\Psi\| .$$

In each summand the first factor is bounded uniformly in $t \in \mathbb{R}$, the second factor is of the form

$$(8.67) \qquad \|C \; e^{-iHt}(H-z)\Psi\|$$

where C is compact by Lemma 8.8 (a). Therefore the time average of these expressions vanishes by Lemma 8.3 as $\tau \to \infty$ because $\Psi \in \mathcal{H}^{cont}(H)$. The second term of (8.65) is estimated the same way. For the first term observe that

$$(8.68) \qquad i[H,D] = i[H_0,D] + \sum_\alpha \; i[V^\alpha,D]$$

$$= 2H_0 + \sum_\alpha i[V^\alpha,D^\alpha],$$

since V^α and d^α commute. The first term on the right-hand side of (8.68) gives the desired result, it remains to estimate the interaction terms. For each α they depend only on the internal variables of the chosen pair and the expressions are analogous to the two-body case (2.16).

$$(8.69) \quad i[V^\alpha,D^\alpha] = -x^\alpha\cdot(\nabla V^\alpha_\ell)(x^\alpha) + iV^\alpha_s \; x^\alpha\cdot p^\alpha - ip^\alpha\cdot x^\alpha \; V^\alpha_s + \nu V^\alpha_s$$

has the property that

$$(8.70) \qquad (h^\alpha-z)^{-1} \; i[V^\alpha,D^\alpha] \; (h^\alpha-z)^{-1}$$

is compact in the 2-body subspace. Writing Q^N as in (8.16) choose $N_1(\varepsilon)$ large enough such that for all $N \geq N_1(\varepsilon)$

$$(8.71) \quad \sum_\alpha\|(h^\alpha-z)^{-1} \; i[V^\alpha,D](h^\alpha-z)^{-1} \sum_{j>N} P^\alpha_j\|\cdot\sup_t\|(h^\alpha-z)e^{-iHt}\Psi\|^2 < \varepsilon/8.$$

Then for this part the time average is bounded by

$$(8.72)\qquad \sup_t \sum_\alpha |(\Psi, e^{iHt}[P^{cont}(h^\alpha) + \sum_{j>N} P_j^\alpha]\ i[V^\alpha, D^\alpha] \sum_{j>N} P_j^\alpha\ e^{-iHt}\Psi)|$$

$$\le \varepsilon/8.$$

For any finite N, $\beta \neq \alpha$

$$(8.73)\qquad \sum_{j\le N} (H-z)^{-1}\ i[V^\alpha, D^\alpha]\ P_j^\beta\ (H-z)^{-1}$$

and

$$(8.74)\qquad \sum_{j\le N} (H-z)^{-1}\ P_j^\beta\ i[V^\alpha, D^\alpha]\ (H-z)^{-1}$$

are compact and the time averages vanish as $\tau \to \infty$ for the corresponding terms. For suitably chosen $r = r(\varepsilon)$

$$(8.75)\qquad \sum_\alpha \|(h^\alpha - z)^{-1}\ i[V^\alpha, D^\alpha](h^\alpha - z)^{-1}\ F(|x^\alpha|>r)\ P^{cont}(h^\alpha)\| \cdot$$

$$\cdot\ \|(h^\alpha - z)e^{-iHt}\Psi\|^2$$

$$< \varepsilon/8.$$

The last remaining term is then bounded by

$$(8.76)\qquad \text{const}\ \|F(|x^\alpha|<r)\ (h^\alpha - z)\ P^{cont}(h^\alpha)\ e^{-iHt}\Psi\|.$$

Its time average vanishes as $\tau \to \infty$ by Proposition 8.4. Summing up the estimates gives (8.63). ■

<u>Proposition</u> <u>8.10</u>. Let H be as in Theorem 8.1. For any $\varepsilon > 0$, $\Psi \in \mathfrak{D}$ (see (8.35)) there is an $N'(\varepsilon)$ such that for all $N \geq N'(\epsilon)$, Q^N as given in (8.1),

$$(8.77)\quad \limsup_{\tau\to\infty} \Big[\frac{\mu^\alpha}{2}\|(\frac{x^\alpha}{\tau} - \frac{p^\alpha}{\mu^\alpha})\, Q^N\, e^{-iH\tau}\Psi\|^2 + \frac{\nu^\alpha}{2}\|(\frac{y^\alpha}{\tau} - \frac{q^\alpha}{\nu^\alpha})\, Q^N\, e^{-iH\tau}\Psi\|^2$$

$$- G_0(\tau) + \frac{2}{\tau^2}\int_0^\tau dt\ t\ G_0(t)\Big] < 3\varepsilon/8,$$

where we use the shorthand

$$(8.78)\quad G_0(t) := (\Psi, e^{iHt}\, Q^N\, H_0\, Q^N\, e^{-iHt}\Psi).$$

<u>Proof</u>.

$$(8.79)\quad \Big|\frac{1}{\tau^2}\int_0^\tau dt\ (\Psi, e^{iHt}\, Q^N\, D\, Q^N\, e^{-iHt}\Psi)$$

$$- \frac{1}{\tau^2}\int_0^\tau dt \int_0^t ds\ 2G_0(s)\Big|$$

$$< \frac{1}{\tau^2}\int_0^\tau dt\ t \cdot \varepsilon/4 = \varepsilon/8$$

for τ large enough by Lemma 8.9. Combining (8.48) with (8.63) and (8.79) gives

$$(8.80)\quad \limsup_{\tau\to\infty} \Big|(\Psi, e^{iH\tau}\, Q^N\, \{\frac{1}{2}\frac{X^2}{\tau^2} - \frac{D}{\tau} + H_0\}\, Q^N\, e^{-iH\tau}\Psi)$$

$$- \{\frac{1}{\tau^2}\int_0^\tau dt \int_0^t ds\ 2G_0(s) - \frac{1}{\tau}\int_0^\tau dt\ 2G_0(t) + G_0(\tau)\}\Big|$$

$$< 3\varepsilon/8.$$

The first term in this expression equals

$$(8.81)\quad \frac{\mu^\alpha}{2}\|(\frac{x^\alpha}{\tau} - \frac{p^\alpha}{\mu^\alpha})\, Q^N\, e^{-iH\tau}\Psi\|^2 + \frac{\nu^\alpha}{2}\|(\frac{y^\alpha}{\tau} - \frac{q^\alpha}{\nu^\alpha})\, Q^N\, e^{-iH\tau}\Psi\|^2$$

and by partial integration

$$\frac{2}{\tau}\int_0^\tau dt\ G_0(t) - \frac{2}{\tau^2}\int_0^\tau dt \int_0^t ds\ G_0(s) = \frac{2}{\tau^2}\int_0^\tau dt\ t\ G_0(t). \tag{8.82}$$

This completes the proof of the proposition. ■

Now we proceed similarly with the parts of the state where at time τ a pair is in a bound state.

<u>Lemma</u> <u>8.11</u>. Let H be as in Theorem 8.1 and $\Psi \in \mathfrak{D}$. Then

$$\lim_{\tau\to\infty}\{(\Psi, e^{iH\tau}\ P_j^\alpha\ \frac{\nu^\alpha}{2}\ \frac{|y^\alpha|^2}{\tau^2}\ e^{-iH\tau}\Psi) \tag{8.83}$$

$$- \frac{1}{\tau^2}\int_0^\tau dt\ (\Psi, e^{iHt}\ P_j^\alpha\ d^\alpha\ e^{-iHt}\Psi)\} = 0$$

<u>Proof</u>. The proof is analogous to Lemma 8.8 (b).

$$i[H,\ P_j^\alpha\ \frac{\nu^\alpha}{2}|y^\alpha|^2] \tag{8.84}$$

$$= i[H^\alpha, P_j^\alpha\ \frac{\nu^\alpha}{2}|y^\alpha|^2] + \sum_{\beta\neq\alpha}\frac{\nu^\alpha}{2}\ i[V^\beta, P_j^\alpha|y^\alpha|^2]$$

$$= P_j^\alpha\ d^\alpha + \sum_{\beta\neq\alpha}\frac{\nu^\alpha}{2}\ i[V^\beta, P_j^\alpha]\ |y^\alpha|^2.$$

The first contribution is the desired expression. By the proof of Lemma 8.8 (a)

$$(H-z)^{-1}\ i[V^\beta, P_j^\alpha]\ |y^\alpha| \tag{8.85}$$

is compact. With the uniform boundedness of

$$\sup_{0\le t\le\tau}\frac{1}{\tau}\ \||y^\alpha|\ e^{-iHt}\Psi\| \tag{8.86}$$

the time average of the other factor vanishes. ■

Lemma 8.12. For $\Psi \in \mathfrak{D}$

$$(8.87) \qquad \lim_{\tau\to\infty}\{(\Psi, e^{iH\tau} P_j^\alpha \frac{d^\alpha}{\tau} e^{-iH\tau}\Psi)$$

$$- \frac{2}{\tau}\int_0^\tau dt\ (\Psi, e^{iHt} P_j^\alpha k_0^\alpha e^{-iHt}\Psi)\} = 0.$$

Proof.

$$(8.88) \qquad i[H, P_j^\alpha d^\alpha] = i[H^\alpha, P_j^\alpha d^\alpha] + \sum_{\beta\neq\alpha} i[V^\beta, P_j^\alpha d^\alpha]$$

$$= 2P_j^\alpha k_0^\alpha + \sum_{\beta\neq\alpha}\{i[V^\beta, P_j^\alpha]d^\alpha + P_j^\alpha\ i[V^\beta, d^\alpha]\}.$$

For the second term observe that $\| |q^\alpha| \exp(-iHt)\Psi\|$ is uniformly bounded and

$$(8.89) \qquad (H-z)^{-1}\ i[V^\beta, P_j^\alpha]\ (1+|y^\alpha|)$$

is compact. The last term can be written with the signs as in (7.3) as

$$(8.90) \qquad P_j^\alpha\ i\ V_s^\beta\ d^\alpha - P_j^\alpha\ i\ d^\alpha\ V_s^\beta \mp P_j^\alpha\ x^\beta\cdot\nabla V_\ell^\beta(x^\beta)$$

$$\pm P_j^\beta\ (x^\beta \mp y^\alpha)\cdot\nabla V_\ell^\beta(x^\beta).$$

All terms are compact when multiplied from both sides with $(H-z)^{-1}$. Therefore they do not contribute to the long time average and (8.87) follows. ■

Proposition 8.13. Let H be as in Theorem 8.1 and $\Psi \in \mathfrak{D}$. Then for any α and j

$$(8.91) \qquad \lim_{\tau\to\infty}\{\frac{\nu^\alpha}{2}\|(\frac{y^\alpha}{\tau} - \frac{q^\alpha}{\nu^\alpha})\ P_j^\alpha\ e^{-iHt}\Psi\|^2 - G_j^\alpha(\tau) + \frac{2}{\tau^2}\int_0^\tau dt\ t\ G_j^\alpha(t)\} = 0,$$

where we use the shorthand

(8.92) $$G_j^\alpha(t) := (\Psi, e^{iHt} P_j^\alpha k_0^\alpha e^{-iHt}\Psi).$$

The proof is analogous to that of Proposition 8.10 and we omit it here.

Corollary 8.14. For any $N \geq N'(\varepsilon)$ set

(8.93) $$G(t) := G_0(t) + \sum_\alpha \sum_{j\leq N} G_j^\alpha(t).$$

If there is an arbitrarily large τ such that

(8.94) $$\frac{1}{T}\int_\tau^{\tau+T} dt \ \{G(t) - \frac{2}{t^2}\int_0^t ds\ s\ G(s)\} < \varepsilon/2h$$

then the assumption of Lemma 8.7 holds (and Theorem 8.1 is true).

Proof. By Proposition 8.10 and 8.13 for any given $N \geq N'(\varepsilon)$ and sufficiently large τ

(8.95) $$\frac{1}{T}\int_\tau^{\tau+T} dt \ \{\frac{\mu^\alpha}{2}\|(\frac{x^\alpha}{t} - \frac{p^\alpha}{\mu^\alpha})\ Q^N\ e^{-iHt}\ \Psi\|^2$$
$$+ \frac{\nu^\alpha}{2}\|(\frac{y^\alpha}{t} - \frac{q^\alpha}{\nu^\alpha})\ Q^N\ e^{-iHt}\Psi\|^2$$
$$+ \sum_\alpha \sum_{j\leq N} \frac{\nu^\alpha}{2}\|(\frac{y^\alpha}{t} - \frac{q^\alpha}{\nu^\alpha})\ P_j^\alpha\ e^{-iHt}\Psi\|^2$$
$$- [G(t) - \frac{2}{t^2}\int_0^t ds\ s\ G(s)]\} < \frac{\varepsilon}{2}.$$

(8.94) and (8.95) imply (8.35). ■

Observe that $G(t)$ is a uniformly bounded continuous function and that

(8.96) $$H(t) := \frac{2}{t^2}\int_0^t ds\ s\ G(s)$$

is bounded and continuously differentiable for $t \neq 0$. Moreover

$$H'(t) = \frac{1}{t}[G(t) - \frac{2}{t^2}\int_0^t ds\ s\ G(s)]. \tag{8.97}$$

Thus the following abstract lemma shows (8.94) and thus completes the proof of Theorem 8.1.

Lemma 8.15. Let $H(t) \in C^1(\mathbb{R}_+)$ be a bounded function with

$$\lim_{t\to\infty} |H'(t)| = 0. \tag{8.98}$$

Then for any $0 < T < \infty$ there is a sequence $\tau_n \to \infty$ such that

$$\lim_{n\to\infty} \frac{1}{T}\int_{\tau_n}^{\tau_n+T} dt\ t\ H'(t) = 0. \tag{8.99}$$

Proof. Assume the contrary. There is an $\varepsilon > 0$ and a $\mathfrak{J}(\varepsilon,T)$ such that (since the function is continuous in τ) for all $\tau \geq \mathfrak{J}(\varepsilon,T)$

$$\frac{1}{T}\int_{\tau}^{\tau+T} dt\ t\ H'(t) > 2\varepsilon\ (\text{or} < -2\varepsilon).$$

For any interval $\tau \leq t \leq \tau + T$ one can decompose

$$H'(t) = H_1(t;\tau) + H_2(t;\tau)$$

where

$$|H_1(t;\tau)| \leq |H'(t)|,$$

$$\int_{\tau}^{\tau+T} dt\ H_1(t;\tau) = 0,$$

$$H_2(t;\tau) = 0 \quad \text{if} \quad H(\tau + T) - H(\tau) = 0,$$

$$\text{sign}[H(\tau+T)-H(\tau)] \cdot H_2(t,\tau) \geq 0 \text{ otherwise}.$$

It follows that

$$H(\tau + T) - H(\tau) = \int_{\tau}^{\tau+T} dt\ H_2(t;\tau).$$

$$\left|\frac{1}{T}\int_{\tau}^{\tau+T} dt\ t\ H_1(t;\tau)\right| = \left|\frac{1}{T}\int_{\tau}^{\tau+T} dt\ (t-\tau)\ H_1(t;\tau)\right|$$

$$\leq T \sup_{\tau\leq t\leq \tau+T} |H_1(t;\tau)| \leq T \sup_{\tau\leq t}|H'(t)| \to 0$$

as $\tau \to \infty$ by (8.98). Thus for all sufficiently large τ

$$\frac{1}{T}\int_{\tau}^{\tau+T} dt\ t\ H_2(t;\tau) > \varepsilon$$

which implies in particular $H_2(t;\tau) \geqslant 0$ and

$$\frac{1}{T}\int_{\tau}^{\tau+T} dt\ (\tau + T)\ H_2(t;\tau) = \frac{T+\tau}{T}\ [H(\tau+T)-H(\tau)] > \varepsilon.$$

Thus for sufficiently large τ

$$H(\tau + T) - H(\tau) > \varepsilon T/(\tau + T)$$

$$H(nT) - H(mT) = \sum_{k=m+1}^{n} H(kT) - H((k-1)T)$$

$$> \varepsilon \sum_{k=m+1}^{n} \frac{1}{k}.$$

For any m this diverges as $n \to \infty$, in contradiction to the boundedness of H. ■

IX. Phase Space Localization of Scattering States.

When the two-body potentials decay towards infinity then the negative spectrum of each h^{α} is discrete. By the HVZ-theorem

$$\Sigma := \min_{\alpha} \inf \sigma(h^{\alpha}) = \inf \sigma^{cont}(H) > -\infty. \tag{9.1}$$

The set of thresholds

$$(9.2) \qquad \mathfrak{J} := \bigcup_{\alpha} \sigma^{pp}(h^{\alpha}) \cup \{0\}$$

is a closed countable set with 0 as the only possible accumulation point. Let $I(E_1,E_4)$ be the subset of the interval $[\Sigma,E_4]$ where an E_1-neighborhood is omitted around each threshold value. Only finitely many intervals have to be removed from $[\Sigma,E_4]$. Then

$$(9.3) \qquad \bigcup_{0<E_1<E_4<\infty} \{\Psi \in \mathcal{H}^{cont}(H) |\ F(H\epsilon I(E_1,E_4))\Psi = \Psi\}$$

is dense in $\mathcal{H}^{cont}(H)$. We will study in the sequel an arbitrary fixed state Ψ from this dense set. We set

$$(9.4) \qquad E_2 := E_4 - \Sigma \geq E_4.$$

Then $H_0 \in I(E_1,E_4)$ implies

$$(9.5) \qquad 0 < E_1 \leq \frac{(p^{\alpha})^2}{2\mu^{\alpha}} + \frac{(q^{\alpha})^2}{2\nu^{\alpha}} \leq E_2 < \infty,$$

and $H^{\alpha} \in I(I_1,E_4)$ implies

$$(9.6) \qquad 0 < E_1 \leq \frac{(q^{\alpha})^2}{2\nu^{\alpha}} \leq E_2 < \infty.$$

We first construct a decomposition of the identity on the subset (9.5) of $\mathbb{R}^{2\nu}$. Let for some $0 < E_3 < E_1$ the function g satisfy

$$(9.7) \qquad g \in C_0^{\infty}(\mathbb{R}),\ (1 - g^2)^{1/2} \in C^{\infty}(\mathbb{R}),\ 0 \leq g \leq 1,$$

$$(9.8) \qquad g(\omega) = 1 \text{ for } 0 \leq \omega \leq E_3,$$

$$(9.9) \qquad g(\omega) = 0 \text{ for } \omega \geq 3E_3/2.$$

For sufficiently small E_3 the subsets of (9.5) which in addition satisfy $(p^{\alpha})^2/2\mu^{\alpha} < 2E_3$ are pairwise disjoint. The maximal speed v of the interior motion for all three pairs is by Theorem 6.1

(9.10) $$v = \max_{\alpha}(4E_3/\mu^{\alpha})^{1/2}.$$

We choose E_3 and thus v small enough such that the minimal speed of any particle relative to the center of mass of the other two is bounded below by 5v:

(9.11) $$\min_{\alpha} \frac{|q^{\alpha}|}{\nu^{\alpha}} = \min_{\alpha}[2(E_1-2E_3)/\nu^{\alpha}]^{1/2} \geq 5v > 0.$$

Due to (9.5) the momenta q^{α} are bounded above as well. Thus there is a finite decomposition of the identity like in Section II of functions $f_i \in C_0^{\infty}(\mathbb{R}^{\nu})$, $0 \leq f_i(w) \leq 1$, such that for a suitable finite collection of $w_i \in \mathbb{R}^{\nu}$, $|w_i| \geq 5v$

(9.12) $$\operatorname{supp} f_i \subset \{w \in \mathbb{R}^{\nu} |\ |w-w_i| < v\},$$

(9.13) $$\sum_i f_i^2(w) = 1 \text{ in a neighborhood of } 5v \leq |w| \leq \max_{\alpha}[2E_2/\nu^{\alpha}]^{1/2}.$$

Then for all α and (p^{α},q^{α}) which satisfy (9.5):

(9.14) $$g((p^{\alpha})^2/2\mu^{\alpha}) = g((p^{\alpha})^2/2\mu^{\alpha}) \sum_i f_i^2(q^{\alpha}/\nu^{\alpha}).$$

It remains to consider the set of points in (9.5) with

(9.15) $$\{(v^{\alpha},w^{\alpha}) \in \mathbb{R}^{\nu}\times\mathbb{R}^{\nu} |\ \tfrac{\mu^{\beta}}{2}(v^{\beta})^2 \geq E_3,\ \tfrac{\mu^{\beta}}{2}(v^{\beta})^2 + \tfrac{\nu^{\beta}}{2}(w^{\beta})^2 \leq E_2,\ \text{all } \beta\}.$$

Note that for any β $v^{\beta} = p^{\beta}/\mu^{\beta}$ and $w^{\beta} = q^{\beta}/\nu^{\beta}$ are explicitly known linear functions of v^{α} and w^{α}. (The transformation is the same as for the map $(x^{\alpha},y^{\alpha}) \to (x^{\beta},y^{\beta})$.) The compact set (9.15) has a finite open cover by balls in $\mathbb{R}^{2\nu}$ around points $(v_j^{\alpha},w_j^{\alpha})$ in (9.15) with radius

(9.16) $$u := v/6 < \min_{\beta} |v^{\beta}(v_j^{\alpha},w_j^{\alpha})|/4.$$

Corresponding to this open cover there is a smooth finite decomposition of the identity $\bar{f}_j^\alpha \in C_0^\infty(\mathbb{R}^{2\nu})$, $0 \le \bar{f}_j^\alpha(v^\alpha,w^\alpha) \le 1$ which satisfies

$$(9.17) \qquad \operatorname{supp} \bar{f}_j^\alpha \subset \{(v^\alpha,w^\alpha)\in\mathbb{R}^{2\nu} \mid |v^\alpha - v_j^\alpha|^2 + |w^\alpha - w_j^\alpha|^2 < u^2\},$$

$$(9.18) \qquad \sum_j (\bar{f}_j^\alpha)^2 (v^\alpha,w^\alpha) = 1 \text{ on a neighborhood of (9.15).}$$

A change of the coordinate system $(v^\alpha,w^\alpha) \to (v^\beta,w^\beta)$ induces the change of the functions

$$(9.19) \qquad \bar{f}_j^\beta(v^\beta,w^\beta) = \bar{f}_j^\alpha(v^\alpha(v^\beta,w^\beta),w^\alpha(v^\beta,w^\beta)).$$

We denote by $f_j^\alpha(v^\alpha,w^\alpha)$ or $f_j^\alpha(p^\alpha/\mu^\alpha,q^\alpha/\nu^\alpha)$ also the multiplication operator with the function $f_j^\alpha(p^\alpha/\mu^\alpha,q^\alpha/\nu^\alpha)$ (viewed as a function of p^α,q^α) which is applied to the momentum space wave function $\psi^\alpha(p^\alpha,q^\alpha)$ of a state Ψ using the same coordinates labelled by the pairing α. As an operator on the state space it is independent of α due to (9.19).

Finally we set

$$(9.20) \qquad f_j^\alpha(v^\alpha,w^\alpha) := [1 - \sum_\beta g^2((p^\beta)^2/2\mu^\beta)]^{1/2}\ \bar{f}_j^\alpha(v^\alpha,w^\alpha).$$

We have constructed the following finite decomposition of the identity into smooth functions of compact support

$$(9.21) \qquad \sum_\beta [g^2((p^\beta)^2/2\mu^\beta) \sum_i f_i^2(q^\beta/\nu^\beta)]$$

$$+ \sum_j (f_j^\alpha)^2 (p^\alpha/\mu^\alpha,q^\alpha/\nu^\alpha) = 1$$

if (p^α,q^α) satisfy (9.5). As an operator it is independent of α. By the remarks following (9.15) also the multiplication operators in configuration space

$$f_j^{\alpha}(x^{\alpha}/\tau, y^{\alpha}/\tau)$$

are independent of the chosen coordinate system α for all τ. Note in particular that by (9.16),(9.17) for *all* β

(9.22) $$\text{supp } f_j^{\alpha}(v^{\alpha}, w^{\alpha}) \subset \{(v^{\beta}, w^{\beta}) \in \mathbb{R}^{2\nu} \mid |v^{\beta}| > 3u\}.$$

In addition the conditions (9.11) and (9.13) guarantee that the finite sum obeys for any α

(9.23) $$\sum_i f_i^2(q^{\alpha}/\nu^{\alpha}) = 1 \text{ on a neighborhood of (9.6).}$$

As a consequence of (9.12),(9.13) also

(9.24) $$f_i(q^{\alpha}/\nu^{\alpha}) = 0 \text{ if } |q^{\alpha}/\nu^{\alpha}| \le 4v, \text{ all } i, \alpha.$$

Without loss of generality we may assume that the norm of the operators corresponding to the sums (9.21) and (9.23) are bounded by one.

These decompositions of the identity and the results of the previous section are used now to localize scattering states in phase space.

Let $\bar{g} \in C_0^{\infty}(\mathbb{R})$ be one on the set $I(E_1, E_4)$ and vanish outside a small neighborhood of it such that the decompositions (9.21) and (9.23) sum up to one on the sets characterized by supp $g(H_0)$ and $g(H^{\alpha})$, respectively. Then for any given $\varepsilon > 0$ and the finitely many functions $g, \bar{g} \in C_0^{\infty}(\mathbb{R})$, $f_i \in C_0^{\infty}(\mathbb{R}^{\nu})$, $f_j^{\alpha} \in C_0^{\infty}(\mathbb{R}^{2\nu})$ choose $N = N(\varepsilon)$ and $\tau = \tau(\varepsilon)$ such that all the estimates (8.2)-(8.8) of Theorem 8.1 hold for all these functions. Then

(9.25) $$e^{-iH\tau}\psi = \bar{g}(H)e^{-iH\tau}\psi =$$

$$= \bar{g}(H)\ Q^N\ e^{-iH\tau}\Psi + \sum_\alpha \bar{g}(H) \sum_{j\le N} P_j^\alpha\ e^{-iH\tau}\Psi$$

$$\approx \bar{g}(H_0)\ Q^N\ e^{-iH\tau}\Psi + \sum_\alpha \bar{g}(H^\alpha) \sum_{j\le N} P_j^\alpha\ e^{-iH\tau}\Psi$$

$$= \{\sum_\beta [g^2((p^\beta)^2/2\mu^\beta)\sum_i f_i^2(q^\beta/\nu^\beta)]$$

$$+ \sum_j (f_j^\alpha)^2(p^\alpha/\mu^\alpha, q^\alpha/\nu^\alpha)\}\ \bar{g}(H_0)\ Q^N\ e^{-iH\tau}\Psi$$

$$+ \sum_\alpha [\sum_i f_i^2(q^\alpha/\nu^\alpha) \sum_{j\le N} P_j^\alpha\ \bar{g}(H^\alpha)]\ e^{-iH\tau}\Psi.$$

The error between the second and third line is bounded by 2ε due to (8.6) and (8.8). With an additional error of 3ε we may omit the operators $\bar{g}(\cdot)$ in the final expression in (9.25):

$$\sum_\alpha \| \sum_{j\le N} P_j^\alpha\ [\bar{g}(H^\alpha)-\bar{g}(H)]\ e^{-iH\tau}\Psi\| < \varepsilon,$$

$$\|[\bar{g}(H_0)-\bar{g}(H)]\ Q^N\ e^{-iH\tau}\Psi\| < \varepsilon,$$

(9.26) $$[\bar{g}(H), Q^N] = -\sum_\alpha [\bar{g}(H), \sum_{j\le N} P_j^\alpha]$$

$$= -\sum_\alpha [(\bar{g}(H)-\bar{g}(H^\alpha)), \sum_{j\le N} P_j^\alpha]$$

$$= \sum_\alpha \sum_{j\le N} \{P_j^\alpha\ [\bar{g}(H)-\bar{g}(H^\alpha)] - [\bar{g}(H)-\bar{g}(H^\alpha)]\ P_j^\alpha\},$$

(9.27) $$\|[\bar{g}(H), Q^N]\ e^{-iH\tau}\Psi\| < \varepsilon$$

by (8.8) and (8.6). Now we apply (8.2)-(8.4) and (8.7) to get as a further approximation of the r.h.s. of (9.25)

$$(9.28)\qquad \{\sum_{\beta} [g(h_0^{\beta})g(\mu^{\beta}(x^{\beta})^2/2\tau^2) \sum_i f_i(q^{\beta}/\nu^{\beta})\ f_i(y^{\beta}/\tau)]$$

$$+ \sum_j f_j^{\alpha}(p^{\alpha}/\mu^{\alpha}, q^{\alpha}/\nu^{\alpha}) f_j^{\alpha}(x^{\alpha}/\tau, y^{\alpha}/\tau)\}\ Q^N\ e^{-iH\tau}\Psi$$

$$+ \sum_{\alpha}[\sum_i f_i(q^{\alpha}/\nu^{\alpha})\ f_i(y^{\alpha}/\tau) \sum_{j\le N} P_j^{\alpha}]\ e^{-iH\tau}\Psi.$$

If there are K terms in the decompositions (9.21) and (9.23) then the error is bounded by $K\varepsilon$. Note that K is independent of ε and $N = N(\varepsilon)$. For the first term in (9.28) observe that the g-terms commute with the f_i's and that for large enough τ

$$(9.29)\qquad \|[g(h_0^{\beta}), g(\mu^{\beta}(x^{\beta})^2/2\tau^2)]\| + \|[g(h^{\beta}), g(\mu^{\beta}(x^{\beta})^2/2\tau^2)]\| < \varepsilon$$

by (6.19). Thus we can apply (8.5) and with an error of 6ε replace $g(h_0^{\beta})$ by $g(h^{\beta})$. Since ε was arbitrary we have shown the following

<u>Proposition</u> <u>9.1</u>. Let H be as in Theorem 8.1 and let for some $0 < E_1 < E_4 < \infty$ $\quad F(H \in I(E_1, E_4))\Psi = \Psi \in \mathcal{H}^{cont}(H)$. (i.e. the energy support of Ψ is bounded above by E_4 and it is separated by E_1 from the thresholds.) For the finite decompositions (9.21) and (9.23) as constructed above and any $\varepsilon > 0$ there is an $N = N(\varepsilon)$ and an arbitrary large $\tau = \tau(\varepsilon)$ such that

$$(9.30)\qquad \|e^{-iH\tau}\Psi - \{ \sum_{\beta} [g(h^{\beta})g(\mu^{\beta}(x^{\beta})^2/2\tau^2) \sum_i f_i(q^{\beta}/\nu^{\beta})f_i(y^{\beta}/\tau)]$$

$$+ \sum_j f_j^{\alpha}(p^{\alpha}/\mu^{\alpha}, q^{\alpha}/\nu^{\alpha})\ f_j^{\alpha}(x^{\alpha}/\tau, y^{\alpha}/\tau)\}\ Q^N\ e^{-iH\tau}\Psi$$

$$- \sum_{\alpha} [\sum_i f_i(q^{\alpha}/\nu^{\alpha})\ f_i(y^{\alpha}/\tau) \sum_{j\le N} P_j^{\alpha}]\ e^{-iH\tau}\Psi\| < \varepsilon.$$

Q^N was defined in (8.1), the expression in braces is independent of α.

We have shown that an "old" scattering state can be decomposed into pieces with the following phase space characteristics: (i) The internal (kinetic) energy of a pair is small and the particles are not too far separated or the two particles are in a bound state. The third particle is far away from both particles in the pair and it has a relatively high velocity pointing away from the pair. We will show below that the third particle will not have a significant interaction with the pair in the future. (ii) All particles are pairwise far separated and they are outgoing relative to each other. There won't be a significant interaction between any of the particles in the future.

We obtained these results without any detailed knowledge of the interacting time evolution exp(-iHt). We used in Section VIII that the short-range potential and the gradient of the long-range potential decay faster than $(1 + |x|)^{-1}$. This was used to obtain compact operators in the whole Hilbert space or in the two body subspaces for all expressions involving the interactions. The elementary abstract fact that the long time average of a compact operator vanishes on the continuous spectral subspace (2.27) and its extension, Proposition 8.4, are sufficient to show that certain interactions asymptotically do not affect the motion. It is remarkable that these simple arguments are sufficient to prove Proposition 9.1. The length of the proof mainly comes from the facts that various operators commute only asymptotically and from the complicated kinematics of three-body systems.

We have used here the implicit technical assumption (7.23) that all two-body bound states have suitable decay in space. Except for zero eigenvalues it is known that this is not an additional assumption [18]. We are convinced, however, that this assumption can be avoided by using additional cutoffs.

X. Three-Body Completeness for Short-Range Potentials.

The wave operators for three-body short-range scatering are

$$(10.1) \qquad \Omega_{\pm}^{0} := \underset{t\to\pm\infty}{\text{s-lim}}\; e^{iHt}\, e^{-iH_0 t}$$

$$(10.2) \qquad \Omega_{\pm,j}^{\alpha} := \underset{t\to\pm\infty}{\text{s-lim}}\; e^{iHt}\, e^{-iH^{\alpha}t}\, P_j^{\alpha}$$

$$(10.3) \qquad \Omega_{\pm}^{\alpha} := \underset{t\to\pm\infty}{\text{s-lim}}\; e^{iHt}\, e^{-iH^{\alpha}t}\, P^{pp}(h^{\alpha})$$

$$\equiv \bigoplus_j \Omega_{\pm,j}^{\alpha}$$

In dimensions $\nu \geqslant 3$ the existence of the strong limits is well known for a large class of potentials. For arbitrary dimension one either needs the eigenfunction decay assumption (7.23) to prove existence (10.2), or in [13] we gave an existence proof without any implicit condition (using that for $E_j^{\alpha} > 0$ (7.23) is satisfied by [18]). If the wave operators exist then it is easy to show that their ranges are pairwise orthogonal and by the intertwining properties

$$(10.4) \qquad H\,\Omega_{\pm}^{0} = \Omega_{\pm}^{0}\, H_0$$

$$(10.5) \qquad H\,\Omega_{\pm}^{\alpha} = \Omega_{\pm}^{\alpha}\, H^{\alpha}$$

$$(10.6) \qquad H\,\Omega_{\pm,j}^{\alpha} = \Omega_{\pm,j}^{\alpha}\, H^{\alpha} = \Omega_{\pm,j}^{\alpha}(k_0^{\alpha} + E_j^{\alpha})$$

the ranges all lie in the absolutely continuous spectral subspace $\mathcal{H}^{ac}(H)$.

The statement of *asymptotic completeness* is

$$(10.7) \qquad \mathcal{H}^{cont}(H) = \mathcal{H}^{ac}(H) = \text{Ran}(\Omega_{\pm}^{0}) \oplus \bigoplus_{\alpha} \text{Ran}(\Omega_{\pm}^{\alpha})$$

for both signs separately. An equivalent statement is (analogous to the two-body case, see Section III):

For any $\Psi \in \mathcal{H}^{cont}(H)$ there is an orthogonal decomposition

$$\Psi = \Psi^0 + \sum_\alpha \Psi^\alpha \tag{10.8}$$

such that for positive times (similarly for negative times)

$$\lim_{\tau\to\infty} \sup_{t\geq 0} \|(e^{-iH_0t} - e^{-iHt})\, e^{-iH\tau}\, \Psi^0\| = 0, \tag{10.9}$$

$$\lim_{\tau\to\infty} \sup_{t\geq 0} \|(e^{-iH^\alpha t} - e^{-iHt})\, e^{-iH\tau}\, \Psi^\alpha\| = 0; \tag{10.10}$$

in addition one has

$$\lim_{\tau\to\infty} \|[\mathbb{1} - P^{pp}(h^\alpha)]\, e^{-iH\tau}\, \Psi^\alpha\| = 0. \tag{10.11}$$

Using two-body completeness one can give a weaker criterion which implies asymptotic completeness (see Lemma 3.3 in [10]):

For any vector Ψ from a dense set in $\mathcal{H}^{cont}(H)$ and any $\varepsilon > 0$ there is a decomposition (not necessarily orthogonal) such that

$$\|\Psi - \tilde{\Psi}^0 - \sum_\alpha \tilde{\Psi}^\alpha\| < \varepsilon \tag{10.12}$$

and there is a $\tau = \tau(\varepsilon)$ such that

$$\sup_{t\geq 0} \|(e^{-iH_0t} - e^{-iHt})\, e^{-iH\tau}\, \tilde{\Psi}^0\| < \varepsilon, \tag{10.13}$$

$$\sup_{t\geq 0} \|(e^{-iH^\alpha t} - e^{-iHt})\, e^{-iH\tau}\, \tilde{\Psi}^\alpha\| < \varepsilon. \tag{10.14}$$

We have to show that $\tilde{\Psi}^{\alpha} \approx \tilde{\tilde{\Psi}}^{0} + \Psi^{\alpha}$ where $\tilde{\tilde{\Psi}}^{0}$ satisfies (10.13):

$$(10.15) \qquad e^{-iH(t+\tau)}\, \tilde{\Psi}^{\alpha} \approx e^{-iH^{\alpha}t}\, [P^{pp}(h^{\alpha})+P^{cont}(h^{\alpha})]\, e^{-iH\tau}\, \tilde{\Psi}^{\alpha}$$

$$=: e^{-iH^{\alpha}t}\, e^{-iH\tau}\, (\Psi^{\alpha} + \tilde{\tilde{\Psi}}^{0}).$$

By completeness for h^{α} , some estimates below, and

$$(10.16) \qquad e^{-iH^{\alpha}\tau'} - e^{-iH_0\tau'} = [e^{-ih^{\alpha}\tau'} - e^{-ih_0^{\alpha}\tau'}]\, e^{-ik_0^{\alpha}\tau'}$$

there is a $\tau' = \tau'(\varepsilon)$ such that

$$(10.17) \qquad \sup_{t\geqslant 0} \| (e^{-iH_0 t} - e^{-iHt})\, e^{-iH(\tau'+\tau)}\, \tilde{\tilde{\Psi}}^{0} \| < \varepsilon .$$

Thus $\tilde{\tilde{\Psi}}^{0}$ belongs to $\mathrm{Ran}(\Omega_{\pm}^{0})$ up to an arbitrarily small error, and Ψ^{α} approximately satisfies (10.10) and (10.11).

The decomposition (10.12) for which we will verify (10.13) and (10.14) is the decomposition into parts with phase space localization properties that was constructed in the last section. The functions used in our phase space decomposition are given in Section IX. Now we will show properties of the time evolutions in the future on the regions of phase space.

Proposition 10.1.. Let f_j^{α} be as characterized in (9.15)-(9.20). Then for $t \cdot \tau > 0$, any β, and $b = 0$ or 1

$$(10.18) \quad \|F(|x^{\beta}|<3u(t+\tau))\, e^{-iH_0 t}\, (h_0^{\beta})^{b}\, f_j^{\alpha}(p^{\alpha}/\mu^{\alpha}, q^{\alpha}/\nu^{\alpha})\, f_j^{\alpha}(x^{\alpha}/\tau, y^{\alpha}/\tau)\|$$

$$\leq C_n (1+|t+\tau|)^{-n} \text{ for all } n \in \mathbb{N}.$$

Proof. Without loss of generality we can choose $\alpha = \beta$ since f_j^{α} can be replaced by f_j^{β} without changing the operators. The norm is bounded by

$$(10.19) \qquad \| F(|x^\alpha - v_j^\alpha (t+\tau)| > u(t+\tau)) \, e^{-ih_0^\alpha t} \, (h_0^\alpha)^b \times$$

$$\times f_j^\alpha (p^\alpha/\mu^\alpha, q^\alpha/\nu^\alpha) \; F(|(x^\alpha/\tau) - v_j^\alpha| < u)\|_{L^2(\mathbb{R}^{2\nu})}$$

$$\leq \sup_{q^\alpha} \| \cdots \|_{L^2(\mathbb{R}^\nu)} .$$

The parameter values q^α can be restricted to the compact set $|q^\alpha/\nu^\alpha - w_j^\alpha| \leq u$ since the operator vanishes otherwise. The rapid decay of (10.19) now follows from Proposition 3.1 for the compact family. ■

Proposition 10.2. Let f_j^α be as defined above and let (7.13) and (7.16) hold for $H = H_0 + \sum_\alpha V_s^\alpha$. Then

$$(10.20) \quad \lim_{\tau\to\infty} \sup_{t\geqslant 0} \| \{ e^{-iHt} - e^{-iH_0 t} \} \, f_j^\alpha (p^\alpha/\mu^\alpha, q^\alpha/\nu^\alpha) \; f_j^\alpha (x^\alpha/\tau, y^\alpha/\tau) \| = 0 .$$

Proof. The supremum is bounded by

$$(10.21) \qquad \int_0^\infty dt \sum_\beta \| V_s^\beta \, e^{-iH_0 t} \, f_j^\alpha \, f_j^\alpha \|$$

$$\leq \int_0^\infty dt \sum_\beta \| V_s^\beta \, (h_0^\beta + \mathbb{1})^{-1} \, F(|x^\beta| > 3u(t+\tau)) \| \, \| (h_0^\beta + \mathbb{1}) \, f_j^\beta \|$$

$$+ \sum_\beta \| V_s^\beta (h_0^\beta + \mathbb{1})^{-1} \| \int_0^\infty dt \; \| F(|x^\beta| < 3u(t+\tau)) \; e^{-iH_0 t} (h_0^\beta + \mathbb{1}) f_j^\beta \; f_j^\beta \| .$$

The first integrand is integrable in $(t + \tau)$ by the short-range condition (7.16). The second decays rapidly by (10.18). This implies (10.20). ■

Proposition 10.3. Assume that (9.7)-(9.13) holds for g and f_i and let $H = H_0 + \sum_\alpha V_s^\alpha$ satisfy (7.13),(7.16). Then for H^α (7.21) and any α

$$\lim_{\tau\to\infty} \sup_{t\geq 0} \|\{e^{-iHt} - e^{-iH^\alpha t}\}\, g(h^\alpha)\, g(\mu^\alpha(x^\alpha)^2/2\tau^2) \times$$

$$f_i(q^\alpha/\nu^\alpha) f_i(y^\alpha/\tau)\| = 0. \tag{10.22}$$

Proof. The supremum is bounded by

$$\int_0^\infty dt \sum_{\beta\neq\alpha} \|V_s^\beta\, e^{-ih^\alpha t}\, g(h^\alpha) g(\mu^\alpha(x^\alpha)^2/2\tau^2)$$

$$\times\, e^{-ik_0^\alpha t}\, f_i(q^\alpha/\nu^\alpha) f_i(y^\alpha/\tau)\|. \tag{10.23}$$

For each β the integrand is majorized by

$$(\|(h^\alpha+i)g(h^\alpha)\| + \|k_0^\alpha\, f_i(q^\alpha/\nu^\alpha)\|) \times \tag{10.24}$$

$$\times \{\|V_s^\beta\, (H^\alpha+i)^{-1}\, F(|x^\alpha|<v(t+\tau))\, F(|y^\alpha|>4v(t+\tau))\| +$$

$$+ \|V_s^\beta\, (H^\alpha+i)^{-1}\| \cdot \sum_{b=0}^{1} \|F(|x^\alpha|>v(t+\tau)) \times$$

$$e^{-ih^\alpha t}\, (h^\alpha+i)^b\, g(h^\alpha)\, g(\mu^\alpha(x^\alpha)^2/2\tau^2)\|$$

$$+ \|(V_s^\beta(H^\alpha+i)^{-1}\| \cdot \sum_{b=0}^{1} \|F(|y^\alpha|<4v(t+\tau))$$

$$e^{-ik_0^\alpha t}\, (k_0^\alpha)^b\, f_i(q^\alpha/\nu^\alpha)\, f_i(y^\alpha/\tau)\|]\}.$$

Since $|x^\beta| > |y^\alpha| - |x^\alpha| > 3v(t + \tau)$ the first term in braces decays integrably in $(t + \tau)$ by the short-range condition in the form (7.27). The integral of the second decays as $\tau \to \infty$ by Theorem 6.1. The last summand decays rapidly in $(t + \tau)$ by Corollary 3.2. Therefore the integral over t decays in τ which implies (10.22). ■

Note that we have used the propagation properties for the interacting time evolution only for two-body subsystems. It is here where we need Theorem 6.1 for small values of v depending on the lower cutoff $E_1 > 0$ in (9.5). If the particles do not bind then Theorem 6.1 holds for subsystems of higher particle number as well and one can prove completeness for N-body systems as well. On the other hand, if particles can bind, then an extension of Theorem 6.1 holds for sufficiently high velocities v corresponding to higher values of the lower cutoff E_1. Similar estimates then will give completeness of high energy N-body scattering.

Corollary 10.4. Let $H = H_0 + \sum_\alpha V_s^\alpha$ satisfy (7.13) and (7.16) and P_j^α, f_i be given with (7.22), (9.23), $P_j^\alpha h^\alpha = E_j^\alpha P_j^\alpha$ with $E_j^\alpha \leq 0$. Then

$$(10.25) \qquad \lim_{\tau\to\infty} \sup_{t\geq 0} \|\{e^{-iHt} - e^{-iH^\alpha t}\} P_j^\alpha f_i(q^\alpha/\nu^\alpha) f_i(y^\alpha/\tau)\| = 0.$$

Proof. For any $E_j^\alpha \leq 0$ there is a g with (9.7)-(9.11) and $g(E_j^\alpha) = 1$, i.e. $g(h^\alpha)P_j^\alpha = P_j^\alpha$. Then

$$(10.26) \qquad \lim_{\tau\to\infty} \|P_j^\alpha - g(h^\alpha)g(\mu^\alpha (x^\alpha)^2/2\tau^2)P_j^\alpha\| = 0.$$

Proposition 10.3 then implies (10.25). ∎

For bound states of two body subsystems with energies at (or below) the threshold value zero we did not need any decay properties to show (10.25). Actually zero energy bound states may have slow decay. For positive energy bound states (if they exist at all) we have to use their known decay properties.

Proposition 10.5. Let a two-body bound state satisfy (7.23), i.e.

$$(10.27) \qquad \|P_j^\alpha F(|x^\alpha|>r)\| \in L^1(\mathbb{R}_+, dr)$$

and let f_i satisfy (9.23). Then

$$(10.28) \qquad \lim_{\tau\to\infty} \sup_{t\geq 0} \|\{e^{-iHt} - e^{-iH^\alpha t}\} P_j^\alpha f_i(q^\alpha/\nu^\alpha) f_i(y^\alpha/\tau)\| = 0.$$

Proof. The estimate is analogous to (10.24) with

$$\|F(|x^\alpha|>v(t+\tau))\ e^{-ih^\alpha t}\ (h^\alpha+i)^b\ P_j^\alpha\|$$

$$\le (|E_j^\alpha| + 1)^b\ \|F(|x^\alpha|>v(t+\tau))\ P_j^\alpha\|$$

integrable in $t + \tau$. The factor $\|(h^\alpha + i)g(h^\alpha)\|$ is replaced by $(|E_j^\alpha| + 1)$. ■

Our main result for short-range three-particle scattering theory is

Theorem 10.6. Let $H = H_0 + \sum_\alpha V_s^\alpha$ satisfy (7.13) and (7.16) [and let the decay condition (7.23) be fulfilled]. Then the wave operators (10.1)-(10.3) exist and are complete (10.7).

Remark. The condition (7.23) is not necessary, see [13]. We used it in the present proof to show Proposition 9.1.

Proof. We verify conditions (10.12)-(10.14) for $F(H\epsilon I(E_1,E_4))\ \Psi = \Psi \in \mathcal{H}^{cont}(H)$. Set

$$(10.29)\qquad \tilde{\Psi}^0 := e^{iH\tau} \sum_j f_j^\alpha(p^\alpha/\mu^\alpha, g^\alpha/\nu^\alpha)\ f_j^\alpha(x^\alpha/\tau, y^\alpha/\tau)\ Q^N\ e^{-iH\tau}\Psi,$$

$$(10.30)\qquad \tilde{\Psi}^\alpha := e^{iH\tau} \sum_i f_i(q^\alpha/\nu^\alpha) f_i(y^\alpha/\tau)$$

$$\times\ [g(h^\alpha)g(\mu^\alpha(x^\alpha)^2/2\tau^2)\ Q^N + \sum_{j\le N} P_j^\alpha]\ e^{-iH\tau}\ \Psi.$$

By Proposition 9.1 there is an N and an arbitrarily large τ such that (10.12) is satisfied. The sums over i and j are finite. Now for τ large enough by Proposition 10.2

$$\sup_{t\geq 0} \|(e^{-iHt} - e^{-iH_0 t}) \sum_j f_j^\alpha(p^\alpha/\mu^\alpha, q^\alpha/\nu^\alpha) f_j^\alpha(x^\alpha/\tau, y^\alpha/\tau)\| \cdot \|Q^N\| < \varepsilon.$$

Then (10.13) is satisfied. Choose τ large enough such that in addition by Propositions 10.3, 10.5, and Corollary 10.4

$$\sup_{t\geq 0}\|(e^{-iH^{\alpha}t}-e^{-iHt})\sum_{i} f_i(q^{\alpha}/\nu^{\alpha})\, f_i(y^{\alpha}/\tau)\,\times$$

$$\times\,[g(h^{\alpha})g(\mu^{\alpha}(x^{\alpha})^2/2\tau^2)\,Q^N + \sum_{j\leq N} P_j^{\alpha}]\| < \varepsilon,$$

then also (10.14) holds and the theorem is proved. ■

XI. Approximate Time Evolution When Long-Range Forces are Present.

As in Section IV we construct an approximate time evolution which takes into account the continuing influence of the tails of the long-range potential. We have to distinguish between the different possibilities to separate one particle from the others.

The sequence $\tau_k = k^{2\rho}$ and the cutoff function φ satisfy (4.1)-(4.3). The sequence of tail parts of the long-range potentials is

$$V_k^{\alpha}(x^{\alpha}) = V_{\ell}^{\alpha}(x^{\alpha})[1-\varphi(x^{\alpha}/u\tau_k)], \tag{11.1}$$

$$V_k := \sum_{\alpha} V_k^{\alpha}(x^{\alpha}), \tag{11.2}$$

with u as given in (9.16). Then

$$[V_k^{\alpha}(x^{\alpha})-V_{\ell}^{\alpha}(x^{\alpha})]\; F(|x^{\alpha}| \geq u\tau_k) = 0, \tag{11.3}$$

$$\sup_{x^{\alpha}} |(\nabla V_k^{\alpha})(x^{\alpha})| \leq \text{const } \tau_k^{-(\delta+3/2)}, \tag{11.4}$$

$$\sup_{x^{\alpha}} |(\Delta V_k^{\alpha})(x^{\alpha})| \leq \text{const } t_k^{-(\delta+2)}. \tag{11.5}$$

We calculate a few commutators which come up in later estimates.

(11.6)
$$[p_j^{\beta}, \exp\{-i \sum_{\alpha\in A} V_k^{\alpha}(x^{\alpha}) t_k\}]$$

$$= -it_k \exp\{-i \sum_{\alpha\in A} V_k^{\alpha}(x^{\alpha}) t_k\} \cdot \sum_{\alpha\in A} [p_j^{\beta}, V_k^{\alpha}(x^{\alpha})].$$

Since any x^{α} is a linear function of x^{β}, y^{β} we obtain with (11.4) for any index set A

(11.7)
$$\| [p_j^{\beta}, \exp\{-i \sum_{\alpha\in A} V_k^{\alpha}(x^{\alpha}) t_k\}] \|$$

$$\le \text{const } t_k \, \tau_k^{-(\delta+3/2)} \le \text{const } \tau_k^{-(1+\varepsilon)}, \ \varepsilon > 0.$$

The same estimates hold for q^{β}.

(11.8)
$$2\mu^{\alpha}[H_0, \exp\{-iV_k^{\alpha}(x^{\alpha}) t_k\}]$$

$$= 2\mu^{\alpha}[h_0^{\alpha}, \exp\{-iV_k^{\alpha}(x^{\alpha}) t_k\}]$$

$$= 2\sum_j [p_j^{\alpha}, \exp\{-iV_k^{\alpha}(x^{\alpha}) t_k\}] p_j^{\alpha}$$

$$+ \sum_j \exp\{\ \} \, [t_k^2 \, (\nabla_j V_k^{\alpha})^2 (x^{\alpha}) + it_k (\Delta V_k^{\alpha})(x^{\alpha})].$$

This estimate and (11.7) yield with (11.4),(11.5)

(11.9)
$$\| [H_0, \exp\{-it \sum_{\alpha\in A} V_k^{\alpha}(x^{\alpha})\}] (H_0 + \mathbb{1})^{-1} \|$$

$$\le \text{const } \tau_k^{-(1+\varepsilon)}, \ \varepsilon > 0, \text{ for } |t| \le t_k,$$

the index set A may contain all three pairings or part of them.

For $\tau_{n+1} \geq t > \tau_n \geq \tau_m$ we define the approximate time evolutions for the total decomposition

$$(11.10) \qquad U^0(t,\tau_m) := e^{-iV_n(t-\tau_n)} e^{-iH_0(t-\tau_n)} U^0(\tau_n,\tau_m),$$

$$(11.11) \qquad U^0(\tau_n,\tau_m) := \prod_{k=m}^{n-1}{}' [e^{-iV_k t_k} e^{-iH_0 t_k}],$$

and for the pairings α

$$(11.12) \qquad U^\alpha(t,\tau_m) := \exp\{-i \sum_{\beta\neq\alpha} V_n^\beta(t-\tau_n)\} e^{-iH^\alpha(t-\tau_n)} U^\alpha(\tau_n,\tau_m),$$

$$(11.13) \qquad U^\alpha(\tau_n,\tau_m) := \prod_{k=m}^{n-1}{}' [\exp\{-i \sum_{\beta\neq\alpha} V_k^\beta t_k\} e^{-iH^\alpha t_k}].$$

Again Π' denotes the time ordered product with increasing indices from right to left, the empty product is the identity operator.

First we show for the case of the total decomposition that U^0 is a good approximation of the true time evolution on suitable subsets of phase space, closely analogous to Proposition 4.2.

Proposition 11.1. Let $H = H_0 + V_s + V_\ell$ satisfy (7.13)-(7.19) and let f_0^α satisfy (9.17) for some (v_0^α,w_0^α) in (9.15); e.g. it may be any of the f_j^α's of Section 9. Then

$$(11.14) \qquad \lim_{m\to\infty} \sup_{t\geq\tau_m} \|\{e^{-iH(t-\tau_m)} - U^0(t,\tau_m)\} f_0^\alpha(p^\alpha/\mu^\alpha, q^\alpha/\nu^\alpha) \times$$

$$f_0^\alpha(x^\alpha/\tau_m, y^\alpha/\tau_m)\| = 0.$$

Proof. As in Section IV we construct an auxiliary family f_k^α of smooth cutoff functions in space. Let χ_k be the characteristic function of a ball around (v_0^α,w_0^α) with radius

(11.15) $$u + \frac{2}{c} \sum_{j=1}^{k} j^{-\rho}.$$

For $\psi \in C_0^\infty(\mathbb{R}^{2\nu})$, $\psi \geq 0$, $\psi(v,w) = 0$ for $|v|^2 + |w|^2 \geq 1$ and $\int d^\nu v\, d^\nu w\, \psi(v,w) = 1$ let

(11.16) $$f_k^\alpha = c(k+1)^\rho \chi_k * \psi[c(k+1)^\rho \cdot], \; k \geq 1.$$

Then the operators $f_k^\alpha(p^\alpha/\mu^\alpha, q^\alpha/\nu^\alpha)$ and $f_k^\alpha(x^\alpha/\tau, y^\alpha/\tau)$ are independent of α if the transformation rule (9.19) is used. For other coordinate systems the supports will be in ellipses but that does not matter. For all c we have

(11.17) $$f_k^\alpha = 1 \text{ on supp } f_{k-1}^\alpha, \; k = 1,2,\ldots$$

(11.18) $$\sup|(\nabla f_k^\alpha)| \leq \text{const } k^\rho,$$

(11.19) $$\|[g(p^\alpha/\mu^\alpha, q^\alpha/\nu^\alpha), f_k^\alpha(x^\alpha/\tau_k, y^\alpha/\tau_k)]\|$$
$$\leq \text{const } k^\rho/\tau_k = \text{const } k^{-\rho}$$

for any $g \in C_0^\infty(\mathbb{R}^{2\nu})$. If c is chosen large enough then in addition for all k and β

(11.20) $$f_k^\beta(v^\beta, w^\beta) = 0 \text{ if } |v^\beta| \leq 2u.$$

(11.23) $$f_0 \equiv f_0(p^\alpha/\mu^\alpha, q^\alpha/\nu^\alpha), \qquad f_k \equiv f_k^\alpha(x^\alpha/\tau_k, y^\alpha/\tau_k).$$

As in Section IV we obtain for the free time evolution

(11.24) $$\|[1-f_{k+1}]\, e^{-iH_0 t_k} f_0 f_k\| \leq C_n(1+t_k)^{-n},$$

and for any $h \in C^\infty(\mathbb{R}^{2\nu})$, any β

(11.25) $$\sup_{t\geq 0} \|F(|x^\beta| \leq u\tau_k)\, e^{-iH_0 t}\, h(p^\alpha, q^\alpha) f_0 f_k\| \quad \leq$$

$$\leq C_n(1+\tau_k)^{-n}.$$

Now we are prepared to estimate (11.14).

$$f_0^\alpha(p^\alpha/\mu^\alpha, q^\alpha/\nu^\alpha)\; f_0^\alpha(x^\alpha/\tau_m, y^\alpha/\tau_m) \tag{11.26}$$

$$\equiv f_0^\alpha\; f_m\; f_0^\alpha(x^\alpha/\tau_m, y^\alpha/\tau_m).$$

Therefore it is sufficient to study (compare (4.25))

$$U^0(t,\tau_m)\, f_0\, f_m \tag{11.27}$$

$$= e^{-iV_n(t-\tau_n)}\, e^{-iH_0(t-\tau_n)} \sum_{k=1}^{n-1} \prod_{j=k+1}^{n-1}{}' \; e^{-iV_j t_j}\, e^{-iH_0 t_j} \times$$

$$\times A_k \prod_{r=m}^{k-1}{}' \, [e^{-iV_r t_r}\, e^{-iH_0 t_r}\, f_r]$$

$$+ e^{-iV_n(t-\tau_n)}\, e^{-iH_0(t-\tau_n)}\, f_0\, f_n \prod_{r=m}^{n-1}{}' \, [e^{-iV_r t_r}\, e^{-iH_0 t_r}\, f_r],$$

where

$$A_k = e^{-iV_k t_k}\, e^{-iH_0 t_k}\, f_0\, f_k - f_0\, f_{k+1}\, e^{-iV_k t_k}\, e^{-iH_0 t_k}\, f_k. \tag{11.28}$$

The expansion is easily verified by induction.

$$\|A_k\| \leq \|[\mathbb{1}-f_{k+1}]\, e^{-iH_0 t_k}\, f_0\, f_k\| \tag{11.29}$$

$$+ \|[f_{k+1}, f_0]\| + \|[e^{-iV_k t_k}, f_0]\|.$$

The first summand is summable in k by (11.24) and the second by (11.19). Since

$$\|[g(p^\alpha/\mu^\alpha, q^\alpha/\nu^\alpha), h(x^\alpha, y^\alpha)]\| \tag{11.30}$$

$$\leq \text{const}(g) \sum_j \{\|[p_j^\alpha, h(x^\alpha, y^\alpha)]\| + \|[q_j^\alpha, h(x^\alpha, y^\alpha)]\|\}$$

the summability in k of the last summand follows from (11.7) and its counterpart for q^α. Thus we have shown

(11.31)
$$\lim_{m\to\infty} \sup_{t>\tau_m} \| U^0(t,\tau_m) f_0 f_m - e^{-iV_n(t-\tau_n)} e^{-iH_0(t-\tau_n)} f_0 f_n \prod_{r=m}^{n-1}{}' [e^{-iV_r t_r} e^{-iH_0 t_r} f_r] \| = 0.$$

After inserting the cutoffs f_r at times τ_r it is now easy to show that it is a good approximation of the interacting time evolution. By induction one verifies

(11.32)
$$e^{-iH(t-\tau_m)} f_0 f_m = e^{-iH(t-\tau_n)} \prod_{j=m}^{n-1} e^{-iHt_j} f_0 f_m =$$
$$= e^{-iV_n(t-\tau_n)} e^{-iH_0(t-\tau_n)} f_0 f_n \prod_{r=1}^{n-1}{}' [e^{-iV_r t_r} e^{-iH_0 t_r} f_r]$$
$$+ e^{-iH(t-\tau_n)} \sum_{k=m}^{n-1} \prod_{j=k+1}^{n-1} e^{-iHt_j} B_k \prod_{r=m}^{k-1} [\ ]$$
$$+ \{e^{-iH(t-\tau_n)} - e^{-iV_n(t-\tau_n)} e^{-iH_0(t-\tau_n)}\} f_0 f_n \prod_{r=1}^{n-1}{}' [\ ],$$

where

(11.33)
$$B_k = e^{-iHt_k} f_0 f_k - f_0 f_{k+1} e^{-iV_k t_k} e^{-iH_0 t_k} f_k$$
$$= \{e^{-iHt_k} - e^{-iV_k t_k} e^{-iH_0 t_k}\} f_0 f_k + A_k.$$

(11.34)
$$\| B_k - A_k \|$$
$$\le \int_0^{t_k} dt' \, \| \frac{d}{dt'} e^{iHt'} e^{-iV_k t'} e^{-iH_0 t'} f_0 f_k \| \le$$

$$\leq \int_0^{t_k} dt' \{\sum_\alpha \|V_0^\alpha (h_0^\alpha+\mathbb{1})^{-1} e^{-iH_0t'} (h_0^\alpha+\mathbb{1}) f_0f_k\|$$

$$+ \sum_\alpha \|(V_\ell^\alpha-V_k^\alpha) e^{-iH_0t'} f_0f_k\|$$

$$+ \|[H_0, e^{-iV_kt'}](H_0+\mathbb{1})^{-1}\|\cdot\|(H_0+\mathbb{1})f_0\|\}.$$

We know from (11.25) that for any α

$$\|F(|x^\alpha| < u\tau_k) e^{-iH_0t'} (h_0^\alpha+\mathbb{1}) f_0f_k\| \tag{11.35}$$

has rapid decay and the integral is summable in k. By the short-range condition (7.16)

$$|t_k|\cdot\|V_s^\alpha (h_0^\alpha+\mathbb{1})^{-1} F(|x^\alpha|>u\tau_k)\| \tag{11.36}$$

is summable in k. (11.3) and the rapid decay of (11.35) imply summability of the second summand in (11.34). With (11.9) the last summand is bounded by $t_k/\tau_k^{1+\varepsilon}$ which is summable in k as well. Since $(t - \tau_n) \leq t_n$ the same estimates apply to the last term in (11.32) and we have shown

$$\lim_{n\to\infty} \sup_{t\geq\tau_m} \|e^{-iH(t-\tau_m)} f_0f_m -$$

$$- e^{-iV_n(t-\tau_n)} e^{-iH_0(t-\tau_n)} f_0f_n \prod_{r=1}^{n-1}{}' [e^{-iV_rt_r} e^{-iH_0t_r} f_r]\|$$

$$= 0. \tag{11.37}$$

This and (11.31) with (11.26) prove the proposition. ■

Now we turn to the case where one particle is separated from a pair α.

Proposition 11.2. Let $H = H_0 + V_s + V_\ell$ satisfy (7.13)-(7.19). Let g, f_0 be from the set of functions g, f_i as constructed in (9.7)-(9.13) with $|w_0| > 5v$. Then for each α with U^α as given in (11.12),(11.13)

$$(11.38) \qquad \lim_{m\to\infty} \sup_{t\geq\tau_m} \|\{e^{-iH(t-\tau_m)} - U^\alpha(t,\tau_m)\} \times$$

$$\times\, g(h^\alpha) g(\mu^\alpha (x^\alpha)^2 / 2\tau_m^2)\; f_0(q^\alpha/\nu^\alpha) f_0(y^\alpha/\tau_m)\| = 0.$$

Proof. As in Section IV we construct a family of spatial cutoff functions which are smoothed characteristic functions of balls around w_0, $|w_0| > 5v$

$$(11.39) \qquad v + \frac{2}{c} \sum_{j=1}^{k} j^{-\rho} < 3v/2 \text{ for all } k.$$

Compare (4.11)-(4.17). With the shorthands

$$(11.40) \qquad f_k \equiv f_k(y^\alpha/\tau_k), \quad f_0 \equiv f_0(q^\alpha/\nu^\alpha)$$

one has $[h^\alpha, f_k] = 0$, and for any $f \in C_0^\infty(\mathbb{R}^\nu)$

$$(11.41) \qquad \|[f(q^\alpha/\nu^\alpha), f_k]\| \leq \text{const}(g) \cdot k^{-\rho}$$

Then as in Section IV for $b = 0,1$

$$(11.42) \qquad f_0(q^\alpha/\nu^\alpha) f_0(y^\alpha/\tau_m) \equiv f_0\; f_m\; f_0(y^\alpha/\tau_m),$$

$$(11.43) \qquad \|[\mathbb{1} - f_{k+1}]\, e^{-ik_0^\alpha t_k} f_0\; f_k\| \leq C_n(1+t_k)^{-n},$$

$$(11.44) \qquad \|F(|y^\alpha| \leq 3v(\tau_k+s))\, e^{-ik_0^\alpha t} (k_0^\alpha)^b f_0\; f_k\| \leq C_n(1+\tau_k)^{-n}.$$

We construct an analogous sequence g_k of cutoff functions which guarantee that the two particles in the pair are not too far separated. If χ_k denotes the characteristic function of a ball around the origin with radius

(11.45) $$v + \frac{2}{c} \sum_{j=1}^{k} j^{-\rho} < 3v/2,$$

then with ψ as in (4.12) we define

(11.46) $$g_k := c(k+1)^{\rho} \chi_k * \psi[c(k+1)^{\rho} \cdot] \in C_0^{\infty}(\mathbb{R}^{\nu}).$$

With (6.19) one has

(11.47) $$\|[g(h^{\alpha}), g_k(x^{\alpha}/\tau_k)]\| \le \mathrm{const}(g) \cdot k^{-\rho},$$

(11.48) $$g_k(x^{\alpha}/\tau_k)\; g(\mu^{\alpha}(x^{\alpha})^2/2\tau_k^2) = g(\mu^{\alpha}(x^{\alpha})^2/2\tau_k^2).$$

With the shorthands

(11.49) $$g \equiv g(h^{\alpha}), \quad g_k \equiv g_k(x^{\alpha}/\tau_k),$$

(11.50) $$\sum_{k=1}^{\infty} \|[\mathbb{1} - g_{k+1}]\; e^{-ih^{\alpha}t_k}\; g\; g_k\| < \infty$$

follows from (6.2) of Theorem 6.1 with the support properties of the family $\{g_k\}$ and the (faster than) linear increase of t_k with k. Similarly for b = 0,1

(11.51) $$\sum_{k=1}^{\infty} \|F(|x^{\alpha}| \geqslant 2v(\tau_k + s))\; e^{-ih^{\alpha}t}\; (h^{\alpha})^b\; g\; g_k\| < \infty.$$

As in the earlier cases we insert cutoff functions $f_k g_k$ at time τ_k into U^{α}. We denote

(11.52) $$\bar{V}_k^{\alpha} \equiv \sum_{\beta \ne \alpha} V_k^{\alpha}$$

(11.53) $$U^{\alpha}(t, \tau_m)\; g\; g_m\; f_0\; f_m$$

$$= e^{-i\bar{V}_n^{\alpha}(t-\tau_n)}\, e^{-iH^{\alpha}(t-\tau_n)}\; g\; g_n\; f_0\; f_n \prod_{k=m}^{n-1}{}' [e^{-i\bar{V}_k^{\alpha}t_k}\; e^{-iH^{\alpha}t_k}\; g_k\; f_k] \;+$$

$$+ e^{-i\bar{V}_n^\alpha(t-\tau_n)} e^{-iH^\alpha(t-\tau_n)} \sum_{k=m}^{n-1} \prod_{j=k+1}^{n-1}{}' e^{-i\bar{V}_j^\alpha t_j} e^{-iH^\alpha t_j} \times$$

$$\times A_k \prod_{r=m}^{k-1}{}' [e^{-i\bar{V}_r^\alpha t_r} e^{-iH^\alpha t_r} g_r f_r],$$

where

(11.54)
$$A_k = e^{-i\bar{V}_k^\alpha t_k} e^{-iH^\alpha t_k} g\, g_k\, f_0\, f_k -$$

$$- g\, g_{k+1}\, f_0\, f_{k+1}\, e^{-i\bar{V}_k^\alpha t_k} e^{-iH^\alpha t_k} g_k\, f_k.$$

Both g and f_0 commute with functions of H^α.

(11.55)
$$\|A_k\| \le \|[1\!\!1-g_{k+1}]\, e^{-ih^\alpha t_k} g\, g_k\|$$

$$+ \|[1\!\!1-f_{k+1}]\, e^{-ik_0^\alpha t_k} f_0\, f_k\|$$

$$+ \|[g,g_{k+1}]\| + \|[g,e^{-i\bar{V}_k^\alpha t_k}]\|$$

$$+ \|[f_0,f_{k+1}]\| + \|[f_0,e^{-i\bar{V}_k^\alpha t_k}]\|.$$

All terms are summable in k by (11.50), (11.43), (11.47), (11.41), (6.19) and (11.7).

(11.56)
$$e^{-iH(t-\tau_n)} \prod_{j=m}^{n-1} e^{-iHt_j} g\, g_m\, f_0\, f_m$$

$$= e^{-i\bar{V}_n^\alpha(t-\tau_n)} e^{-iH^\alpha(t-\tau_n)} g\, g_n\, f_0\, f_n \prod_{k=m}^{n-1}{}' [e^{-i\bar{V}_k^\alpha t_k} e^{-iH^\alpha t_k} g_k\, f_k]$$

$$+ e^{-iH(t-\tau_n)} \sum_{k=m}^{n-1} \prod_{j=k+1}^{n-1} e^{-iHt_j} B_k \prod_{r=m}^{k-1}{}' [\;\;] +$$

$$+ \{e^{-iH(t-\tau_n)} - e^{-i\bar{V}^\alpha_n(t-\tau_n)} e^{-iH^\alpha(t-\tau_n)}\} g\, g_n\, f_0\, f_n \prod_{r=m}^{n-1}{}' [\quad],$$

where

$$(11.57)\quad B_k = e^{-iHt_k} g\, g_k\, f_0\, f_k - g\, g_{k+1}\, f_0\, f_{k+1}\, e^{-i\bar{V}^\alpha_k t_k} e^{-iH^\alpha t_k} g_k\, f_k$$

$$= \{e^{-iHt_k} - e^{-i\bar{V}^\alpha_k t_k} e^{-iH^\alpha t_k}\} g\, g_k\, f_0\, f_k + A_k.$$

$$(11.58)\quad \|B_k - A_k\| \le \int_0^{t_k} dt' \{ \sum_{\beta\ne\alpha} \|V^\beta_s\, (H^\alpha - z)^{-1}\, e^{-iH^\alpha t'}\, (H^\alpha - z)\, g\, g_k\, f_0\, f_k\|$$

$$+ \sum_{\beta\ne\alpha} \|(V^\alpha_\ell - V^\beta_k)\, e^{-iH^\alpha t'}\, g\, g_k\, f_0\, f_k\|$$

$$+ \|[H^\alpha, e^{-i\bar{V}^\alpha_k t'}](H^\alpha - z)^{-1}\| \cdot \|(H^\alpha - z)\, g\, f_0\|\}.$$

The first integrands are split according to

$$(11.59)\quad \|V^\beta_s\, (H^\alpha - z)^{-1}\, F(|y^\alpha| > 3v(\tau_k + t'))\, F(|x^\alpha| < 2v(\tau_k + t'))\|$$

$$\cdot \|(H^\alpha - z)\, g\, f_0\|$$

$$+ \|V^\beta_s\, (H^\alpha - z)^{-1}\| \sum_{b=0}^{1} \{\|F(|y^\alpha| < 3v(\tau_k + t'))\, e^{-ik^\alpha_0 t'}\, (k^\alpha_0 - z)^b\, f_0\, f_k\|$$

$$+ \|F(|x^\alpha| > 2v(\tau_k + t'))\, e^{-ih^\alpha t'}\, (h^\alpha - z)^b\, g\, g_k\|\}.$$

With $|x^\beta| > |y^\alpha| - |x^\alpha|$ all terms are integrable in $\tau_k + t'$ by (7.27), (11.44), and (11.51). Similarly for the second sum since $V^\beta_\ell - V^\beta_k$ vanishes for $|x^\beta| \geqslant v\tau_k$. Uniformly in $|t'| \le t_k$ the last integrand is bounded by const $\tau_k^{-(1+\varepsilon)}$, $\varepsilon > 0$, as given in (11.9). Since $t - \tau_n \le t_n$ the same estimates apply to the last summand in (11.56). Thus we have shown that all error terms are summable in k and (11.38) follows.
■

Corollary 11.3. Let H and f_0 be as in Proposition 11.2. Let $h^\alpha P_j^\alpha = E_j^\alpha P_j^\alpha$ with $E_j^\alpha \le 0$. Then

$$(11.60) \quad \lim_{m\to\infty} \sup_{t\geq\tau_m} \|\{e^{-iH(t-\tau_m)} - U^\alpha(t,\tau_m)\} P_j^\alpha f_0(q^\alpha/\nu^\alpha) f_0(y^\alpha/\tau_m)\| = 0.$$

Proof. There is a suitable g with $g(E_j^\alpha) = 1$ and

$$(11.61) \quad \lim_{\tau\to\infty} \|P_j^\alpha - g(h^\alpha)\, g(\mu^\alpha(x^\alpha)^2/2\tau^2)\, P_j^\alpha\| = 0.$$

With (11.38) this implies (11.60). ∎

It is straightforward to show (11.60) for positive energy bound states with (7.23) as well. There is, however, a simpler approximate time evolution for scattering states with a bounded pair. It has the advantage over U^α that the long-range correction terms commute with h^α. The two cluster system can be treated like a two body system. Choose the right sign such that

$$(11.62) \quad x^\beta = \pm\, y^\alpha + \text{const}\ x^\alpha$$

and replace $V_k^\beta \equiv V_k^\beta(x^\beta)$ by $V_k^\beta(\pm y^\alpha)$ to obtain analogous to (11.12), (11.13) for $\tau_{n+1} \geq t > \tau_n \geq \tau_m$:

$$(11.63) \quad U_b^\alpha(t,\tau_m) := \exp\{-i(t-\tau_n) \sum_{\beta\neq\alpha} V_n^\beta(\pm y^\alpha)\}\, e^{-iH^\alpha(t-\tau_n)}\, U_b^\alpha(\tau_n,\tau_m),$$

$$(11.64) \quad U_b^\alpha(\tau_n,\tau_m) := \prod_{k=m}^{n-1}{}' \exp\{-it_k \sum_{\beta\neq\alpha} V_k^\beta(\pm y^\alpha)\}\, e^{-iH^\alpha t_k}.$$

Proposition 11.4. Let H and f_0 be as above and assume that P_j^α has the decay property (7.23). Then

$$(11.65) \quad \lim_{m\to\infty} \sup_{t\geq\tau_m} \|\{e^{-iH(t-\tau_m)} - U_b^\alpha(t,\tau_m)\} P_j^\alpha f_0(q^\alpha/\nu^\alpha) f_0(y^\alpha/\tau_m)\| = 0.$$

<u>**Proof**</u>. We mimic the proof of Proposition 11.2 with several simplifications. Instead of the sequence gg_k one carries along the fixed P_j^α which commutes with $\exp(-ih^\alpha t)$. In the analogue of (11.55) all terms with g's do not occur. As the counterparts of (11.56)-(11.58) one has to estimate instead

$$(11.66) \qquad \| [e^{-iHt_k} - \exp\{-it_k \sum_{\beta\neq\alpha} V_k^\beta(\pm y^\alpha)\}\, e^{-iH^\alpha t_k}]\, P_j^\alpha\, f_0\, f_k \|$$

$$\leq \int_0^{t_k} dt' \{ \sum_{\beta\neq\alpha} \| V_s^\beta P_j^\alpha\, (H^\alpha - z)^{-1}\, e^{-iH^\alpha t'}\, (H^\alpha - z)\, f_0\, f_k \|$$

$$+ \sum_{\beta\neq\alpha} \| [V_\ell^\beta(x^\beta) - V_k^\beta(\pm y^\alpha)]\, P_j^\alpha\, e^{-iH^\alpha t'}\, f_0\, f_k \|$$

$$+ \| [k_0^\alpha, \exp\{-it' \sum_{\beta\neq\alpha} V_k^\beta(\pm y^\alpha)\}]\, (k_0^\alpha + \mathbb{1})^{-1} \| \cdot \| (k_0^\alpha + \mathbb{1}) f_0 \| \}.$$

For the first sum of integrands use that

$$(11.67) \qquad \| V_s^\beta\, P_j^\alpha\, (H^\alpha - z)^{-1}\, F(|y^\alpha| > 3v(t_k + t')) \|$$

decays integrably in $\tau_k + t'$ as a consequence of (7.27) and (7.23). With (11.44) the summability in k of the corresponding term follows. As in the estimate of (11.58) in the second sum one can replace $V_\ell^\beta(x^\beta)$ by $V_k^\beta(x^\beta)$. With

$$(11.68) \qquad \| [V_k^\beta(x^\beta) - V_k^\beta(\pm y^\alpha)]\, P_j^\alpha \|$$

$$\leq \text{const}\ \tau_k^{-(\delta+3/2)} \| |x^\alpha| P_j^\alpha \|$$

the summability of the integrals follows. The last term in (11.66) is a two-body expression which was estimated in (4.35). The proof is finished analogous to that of Proposition 11.2. ■

XII. Asymptotic Time Evolution on Certain Regions of Phase Space, the Long-Range Case.

We study first the case of the total decomposition of the three particle state. It is closely analogous to the two-body case presented in Section V.

Proposition 12.1. Let U^0 be as defined in (11.10), (11.11) and let the assumptions of Proposition 11.1 be satisfied. Then for all β, $\tau \geqslant \tau_m$

$$(12.1) \qquad \| (\frac{x^\beta}{\tau} - \frac{p^\beta}{\mu^\beta}) \, U^0(\tau,\tau_m) \, f_0^\alpha(p^\alpha/\mu^\alpha, q^\alpha/\nu^\alpha) \, f_0^\alpha(x^\alpha/\tau_m, y^\alpha/\tau_m) \|$$

$$\leq \begin{cases} \text{const}(\tau_m)\cdot\tau^{-(\delta+1/2)}, & 0 < \delta < 1/2, \\ \text{const}\,[\tau_m + \ell n(\tau/\tau_m)]/\tau, & \delta = 1/2. \end{cases}$$

Proof. Since the operators f_0^α are independent of α we can set $\alpha = \beta$. With

$$(12.2) \qquad [x^\alpha - (p^\alpha/\mu^\alpha)\tau] \, e^{-iH_0 t} = e^{-iH_0 t} \, [x^\alpha - (p^\alpha/\mu^\alpha)(\tau - t)]$$

it remains to estimate (analogous to Section V)

$$(12.3) \qquad \sum_{j=m}^{n} \| \tau_{j+1} \, [p^\alpha, e^{-iV_j t_j}] \|$$

$$\leq \text{const}\; \tau_n^{(-\delta+1/2)} \text{ or } \ell n(\tau_n/\tau_m)$$

using (11.7). This implies (12.1). ■

The Dollard modified free time evolution is in this case [5]

$$(12.4) \qquad U_D^0(T,\tau) := e^{-iH_0(T-\tau)} \, U_D^{0\prime}(T,\tau),$$

$$(12.5) \qquad U_D^{0\prime}(T,\tau) := \exp\{-i \int_\tau^T dt \sum_\alpha V_\ell^\alpha(tp^\alpha/\mu^\alpha)\}.$$

Let $\bar{f}^\alpha \in C_0^\infty(\mathbb{R}^{2\nu})$, $0 \le \bar{f}^\alpha \le 1$ satisfy

$$\bar{f}^\alpha = 1 \text{ on } \bigcup_k \operatorname{supp} f_k^\alpha , \tag{12.6}$$

$$\bar{f}^\alpha(v^\alpha, w^\alpha) = 0 \text{ in a neighbourhood of } |v^\beta| \le 3u/2, \text{ any } \beta. \tag{12.7}$$

Then we have the following propagation property.

<u>Lemma</u> <u>12.2</u>. Let V_ℓ satisfy (7.19) and $\bar{f}^\alpha$ be as above. Then for $b = 0$ or 1, any β, uniformly in $1 \le \tau \le T$

$$\| F(|x^\beta| \le uT)\, U_D^0(T,\tau)\, (h_0^\beta + \mathbb{1})^b\, \bar{f}^\alpha(p^\alpha/\mu^\alpha, q^\alpha/\nu^\alpha)\, \bar{f}^\alpha(x^\alpha/\tau, y^\alpha/\tau) \| \tag{12.8}$$

$$\le C_n (1+T)^{-n}.$$

<u>Proof</u>. With $\bar{\bar{f}}^\alpha = 1$ on $\operatorname{supp} \bar{f}^\alpha$, $\bar{\bar{f}}^\alpha(v^\alpha, w^\alpha) = 0$ if $|v^\beta| \le u$ for any β we have the bound (setting $\alpha = \beta$ again)

$$\| F(|x^\alpha| \le uT)\, U_D^{0\prime}(T,\tau)\, \bar{\bar{f}}^\alpha(p^\alpha/\mu^\alpha, q^\alpha/\nu^\alpha)\, F(|x^\alpha| \geqslant 3uT/2) \| \tag{12.9}$$

$$+ \| F(|x^\alpha| \le 3uT/2)\, e^{-iH_0(T-\tau)}\, \bar{f}^\alpha(p^\alpha/\mu^\alpha, q^\alpha/\nu^\alpha)\, \bar{f}^\alpha(x^\alpha/\tau, y^\alpha/\tau) \| .$$

By (10.18) the second summand has rapid decay in T uniform in $1 \le \tau \le T$. The rapid decay of the first summand follows as in the proof of Corollary 2.5 in [11]. ■

The estimate of asymptotic observables carries over to U_D^0.

<u>Lemma</u> <u>12.3</u>. For f_0^α, $\bar{f}^\alpha$, U^0, and U_D^0 as given above and $t \geqslant \tau \geqslant \tau_m$, $0 < \delta < 1/2$, any β

$$\| (\frac{x^\beta}{t} - \frac{p^\beta}{\mu^\beta})\, U_D^0(t,\tau)\, \bar{f}^\alpha(p^\alpha/\mu^\alpha, q^\alpha/\nu^\alpha)\, \bar{f}^\alpha(x^\alpha/\tau, y^\alpha/\tau) \times \tag{12.10}$$

$$\times\, U^0(\tau,\tau_m)\, f_0^\alpha(p^\alpha/\mu^\alpha, q^\alpha/\nu^\alpha)\, f_0^\alpha(x^\alpha/\tau_m, y^\alpha/\tau_m) \| \quad \le$$

$$\leq \text{const}(\tau_m) \cdot t^{(-\delta+1/2)} .$$

We omit the proof which is analogous to that of Lemma 5.3 and (5.4). The main result for the totally decomposed case is

Proposition 12.4. Let $H = H_0 + V_s + V_\ell$ satisfy (7.13)-(7.19) and U^0, U_D^0 be defined in (11.10), (11.11) and (12.4), (12.5), respectively. For f_0^α as in Proposition 12.1 and any m

$$\text{(12.11)} \qquad \lim_{\tau\to\infty} \sup_{T\geq\tau} \| \{ e^{-iH(T-\tau)} - U_D^0(T,\tau) \} \, U^0(\tau,\tau_m) \times$$

$$\times \, f_0^\alpha(p^\alpha/\mu^\alpha, q^\alpha/\nu^\alpha) \; f_0^\alpha(x^\alpha/\tau_m, y^\alpha/\tau_m) \| = 0 .$$

Proof. With

$$V^\alpha(t;x^\alpha) = V_\ell^\alpha(x^\alpha)[1-\varphi(x^\alpha/ut)]$$

the proof is exactly the same as for Proposition 5.4 for each pairing α separately. ■

If the pair α is in a bound state we use as the corresponding Dollard modified time evolution [5]

$$\text{(12.12)} \qquad U_D^\alpha(T,\tau) := e^{-iH^\alpha(T-\tau)} \, U_D^{\alpha\prime}(T,\tau),$$

$$\text{(12.13)} \qquad U_D^{\alpha\prime}(T,\tau) := \exp\{-i \int_\tau^T dt \sum_{\beta\neq\alpha} V_\ell^\beta(tq^\alpha/\nu^\alpha)\}.$$

Proposition 12.5. Let H satisfy (7.13)-(7.19) and P_j^α (7.23). With U_b^α defined in (11.63), (11.64) and U_D^α in (12.12), (12.13) and f_0 as in Proposition 11.4:

$$\text{(12.14)} \qquad \lim_{\tau\to\infty} \sup_{T\geq\tau} \| \{ e^{-iH(T-\tau)} - U_D^\alpha(T,\tau) \} \times$$

$$\times\, U_b^\alpha(\tau,\tau_m)\ P_j^\alpha\ f_0(q^\alpha/\nu^\alpha)\ f_0(y^\alpha/\tau_m)\| = 0$$

for any τ_m.

Proof. We know from Proposition 11.4 that $\exp\{-iH(T-\tau)\}$ is well approximated by $U_b^\alpha(T,\tau)$. Then the internal motion of the pair is trivial $\exp\{-iE_j^\alpha(T-\tau)\}$ and the two-body result Proposition 4.2 applied to the motion of the third particle shows that

$$e^{-iH(T-\tau)} \approx e^{-iH'(T-\tau)} \tag{12.15}$$

where

$$H' = h^\alpha + k_0^\alpha + \sum_{\beta\neq\alpha} V_\ell^\beta(\pm y^\alpha). \tag{12.16}$$

Then Proposition 5.4 is exactly the remaining estimate which shows (12.14). ■

Now we introduce a condition on the state Ψ which ensures that it lies in the sum of the ranges of the wave operators. If any $\Psi \in \mathcal{H}^{cont}(H)$ satisfies this condition then asymptotic completeness holds.

Proposition 12.6. Let $H = H_0 + V_s + V_\ell$ satisfy (7.13)-(7.19) and P_j^α (7.23). Let U_D^0 and U_D^α be as defined in (12.4), (12.5), (12.12), and (12.13). For Ψ from the dense set in $\mathcal{H}^{cont}(H)$: $F(H \in I(E_1,E_4))\Psi = \Psi \in \mathcal{H}^{cont}(H)$ for some $0 < E_1 < E_4 < \infty$ assume in addition the following property: For any $\varepsilon > 0$ there are $E_3 = E_3(\varepsilon)$, $N = N(\varepsilon)$ and arbitrarily large $\tau = \tau(\varepsilon)$ such that with the functions g, f_i, f_j^α as constructed in the beginning of Section IX Proposition 9.1 holds and in addition for such τ

$$\sum_\alpha\{\|F(h^\alpha<2E_3)\ P^{cont}(h^\alpha)\ e^{-iH\tau}\Psi\| + \|\sum_{j>N} P_j^\alpha\ e^{-iH\tau}\Psi\|\} < \varepsilon. \tag{12.17}$$

Then there is a decomposition of Ψ with

$$\|\Psi - \tilde{\Psi}^0 - \sum_\alpha \tilde{\Psi}^\alpha\| < 3\varepsilon, \tag{12.18}$$

and for sufficiently large $T = T(\varepsilon)$, all α,

$$(12.19) \qquad \sup_{t\geq T} \|\{e^{-iH(t-T)} - U_D^0(t,T)\}\, e^{-iHT}\, \tilde{\Psi}^0\| < \varepsilon,$$

$$(12.20) \qquad \sup_{t\geq T} \|\{e^{-iH(t-T)} - U_D^{\alpha}(t,T)\}\, e^{-iHT}\, \tilde{\Psi}^{\alpha}\| < \varepsilon.$$

Proof. We use as in (8.16)

$$Q^N = P^{cont}(h^{\beta}) + \sum_{j>N} P_j^{\beta} - \sum_{\alpha\neq\beta} \sum_{j\leq N} P_j^{\alpha}.$$

By assumption (12.17) for suitable sufficiently large τ

$$(12.21) \qquad \sum_{\beta} \|g(h^{\beta}) g(\mu^{\beta}(x^{\beta})^2/2\tau^2) \sum_i f_i(q^{\beta}/\nu^{\beta})\, f_i(y^{\beta}/\tau)\; Q^N\, e^{-iH\tau}\Psi\|$$

$$\leq \sum_{\beta} \sum_{\alpha\neq\beta} \sum_{j\leq N} \|g(\mu^{\beta}(x^{\beta})^2/2\tau^2) \sum_i f_i(y^{\beta}/\tau)\; P_j^{\alpha}\|$$

$$+ \sum_{\beta}\{\| \sum_{j>N} P_j^{\beta}\; e^{-iH\tau}\Psi\| + \|F(h^{\beta}<2E_3)\; P^{cont}(h^{\beta})\; e^{-iH\tau}\Psi\|\}$$

$$+ \sum_{\beta} \|[g(h^{\beta}), g(\mu^{\beta}(x^{\beta})^2/2\tau^2)]\| < 2\varepsilon.$$

With Proposition 9.1 we have for suitable sufficiently large τ

$$(12.22) \qquad \|e^{-iH\tau}\Psi - \sum_j f_j^{\alpha}(p^{\alpha}/\mu^{\alpha}, q^{\alpha}/\nu^{\alpha})\; f_j^{\alpha}(x^{\alpha}/\tau, y^{\alpha}/\tau)\; Q^N\; e^{-iH\tau}\;\Psi$$

$$- \sum_{\alpha} [\sum_i f_i(q^{\alpha}/\nu^{\alpha}) f_i(y^{\alpha}/\tau) \sum_{j\leq N} P_j^{\alpha}]\; e^{-iH\tau}\Psi\| < 3\varepsilon.$$

By Propositions 11.1 and 11.4 we have for large enough τ_m also

$$(12.23) \qquad \sup_{t\geq\tau_m} \|\{e^{-iH(t-\tau_m)} - U^0(t,\tau_m)\} \sum_j f_j^{\alpha}\; f_j^{\alpha}\| < \varepsilon/3,$$

$$(12.24) \qquad \sup_{t\geqslant\tau_m} \|\{e^{-iH(t-\tau_m)} - U_b^\alpha(t,\tau_m)\} \sum_i f_i \; f_i \sum_{j\leq N} P_j^\alpha\| < \varepsilon/3.$$

Note that in Section XI the parameters k, m, etc. need not be integers, it is sufficient that k-m runs through integer values (or a fixed multiple thereof). Thus τ_m can be chosen such that (12.22) holds. For such a τ_m for which (12.22)-(12.24) hold simultaneously we set

$$(12.25) \qquad \tilde{\Psi}^0 := e^{iH\tau_m} \sum_j f_j^\alpha(p^\alpha/\mu^\alpha, q^\alpha/\nu^\alpha) \; f_j^\alpha(x^\alpha/\tau_m, y^\alpha/\tau_m) \; Q^N \; e^{-iH\tau_m} \Psi,$$

$$(12.26) \qquad \tilde{\Psi}^\alpha := e^{iH\tau_m} \sum_i f_i(q^\alpha/\nu^\alpha) \; f_i(y^\alpha/\tau_m) \sum_{j\leq N} P_j^\alpha \; e^{-iH\tau_m} \Psi.$$

Then (12.18) is satisfied by (12.22). By (12.23), (12.24) uniformly in $T \geqslant \tau_m$

$$(12.27) \qquad \sup_{t\geqslant T} \|[e^{-iH(t-T)} - U_D^0(t,T)] \; e^{-iHT} \; \tilde{\Psi}^0\|$$

$$\leq 2\varepsilon/3 + \sup_{t\geqslant T} \|[e^{-iH(t-T)} - U_D^0(t,T)] \; U^0(T,\tau_m) \; e^{-iH\tau_m} \; \tilde{\Psi}^0\|,$$

$$(12.28) \qquad \sup_{t\geqslant T} \|[e^{-iH(t-T)} - U_D^\alpha(t,T)] \; e^{-iHT} \; \tilde{\Psi}^\alpha\|$$

$$\leq 2\varepsilon/3 + \sup_{t\geqslant T} \|[e^{-iH(t-T)} - U_D^\alpha(t,T)] \; U_b^\alpha(T,\tau_m) \; e^{-iH\tau_m} \; \tilde{\Psi}^\alpha\|.$$

The last summands in both terms are smaller than $\varepsilon/3$ for large T by Propositions 12.4 and 12.5, respectively. This proves (12.19) and (12.20). ■

The Dollard modified wave operators for three particle long-range scattering are defined as follows [5]

$$(12.29)\qquad \Omega^{0}_{D\pm} := \underset{t\to\pm\infty}{s\text{-}\lim}\ e^{iHt}\ U^{0}_{D}(t,0)$$

$$(12.30)\qquad \Omega^{\alpha}_{D\pm j} := \underset{t\to\pm\infty}{s\text{-}\lim}\ e^{iHt}\ U^{\alpha}_{D}(t,0)\ P^{\alpha}_{j}$$

$$(12.31)\qquad \Omega^{\alpha}_{D\pm} := \underset{t\to\pm\infty}{s\text{-}\lim}\ e^{iHt}\ U^{\alpha}_{D}(t,0)\ P^{pp}(h^{\alpha})$$

$$\equiv \bigoplus_{j} \Omega^{\alpha}_{D\pm j}.$$

With the estimates given in the proofs of Propositions 12.2-12.5 it is easy to show existence of the modified wave operators and they satisfy the same intertwining relations (10.4)-(10.6) as in the short-range case. In particular the direct sum of their pairwise orthogonal ranges is contained in the absolutely continuous spectral subspace. [5], [31].

Asymptotic completeness means

$$(12.32)\qquad \mathcal{H}^{cont}(H) = \mathcal{H}^{ac}(H) = \operatorname{Ran}\ \Omega^{0}_{D\pm} \oplus \bigoplus_{\alpha} \operatorname{Ran}\ \Omega^{\alpha}_{D\pm}$$

for each sign separately. As in the short-range case it is sufficient to show that for any $\Psi \in \mathcal{H}^{cont}(H)$ and $\varepsilon > 0$ there is a decomposition of Ψ with

$$(12.33)\qquad \|\Psi - \tilde{\Psi}^{0} - \sum_{\alpha} \tilde{\Psi}^{\alpha}\| < \varepsilon$$

and a $T = T(\varepsilon)$ such that for all α

$$(12.34)\qquad \sup_{t\geqslant T} \|[e^{-iH(t-T)} - U^{0}_{D}(t,T)]\ e^{-iHT}\ \tilde{\Psi}^{0}\| < \varepsilon.$$

$$(12.35)\qquad \sup_{t\geqslant T} \|[e^{-iH(t-T)} - U^{\alpha}_{D}(t,T)]\ e^{-iHT}\ \tilde{\Psi}^{\alpha}\| < \varepsilon.$$

In our last proposition we have shown:

Corollary 12.7. Let the assumptions of Proposition 12.6 be satisfied, then Ψ lies in the direct sum of the ranges of the modified wave operators. In particular, if every Ψ with $F(H \in I(E_1,E_4))\Psi = \Psi \in \mathcal{H}^{cont}(H)$ satisfies (12.17) then asymptotic completeness holds.

As a special case we have shown already that asymptotic completeness holds below zero energy, i.e. below the three particle breakup threshold. For a suitable sequence of late times $H \approx H^\alpha = h^\alpha + k_0^\alpha$. $H \le E < 0$ and $k_0^\alpha \geq 0$ imply $h^\alpha \le E$ and (12.17) is satisfied. For states with positive energy we will show in the next section that (12.17) is satisfied.

XIII. Asymptotic Completeness for Three-Body Systems with Long-Range Potentials.

In the previous section we have shown asymptotic completeness if certain parts of a state are arbitrarily small at late times. Now we verify this assumption for a class of long-range potentials which is a bit smaller than the one considered in the rest of the paper.

The approximate time evolution U^α is defined in (11.12), (11.13). For easier control we introduce a further cutoff. Let $\varphi \in C^\infty(\mathbb{R})$, $0 \le \varphi(z) \le 1$ satisfy

(13.1) $\qquad \varphi(z) = 0 \quad \text{if } |z| \le 1,$

(13.2) $\qquad \varphi(z) = 1 \quad \text{if } |z| \geq 2.$

As above for U^α pick $m \in \mathbb{R}$ such that the initial time for the time evolution $\tau_m \equiv m^{2\rho}$. Then the parameter k proceeds in integer steps starting from m. For $\tau_{n+1} \geq t > \tau_n$ set

(13.3) $\qquad \tilde{U}^\alpha(t,\tau_m) :=$

$$\exp\{-i(t-\tau_n)\sum_{\beta\neq\alpha}(V_n^\beta(\pm y^\alpha)+\varphi(\frac{x^\alpha}{\tau_n^{\delta'}})[V_n^\beta(x^\beta)-V_n^\beta(\pm y^\alpha)])\}\times$$

$$\times\, e^{-iH^\alpha(t-\tau_n)}\tilde{U}^\alpha(\tau_n,\tau_m),$$

(13.4) $$\tilde{U}^\alpha(\tau_n,\tau_m) :=$$

$$\prod_{k=m}^{n-1}{}' [\exp\{-it_k\sum_{\beta\neq\alpha}(V_k^\beta(\pm y^\alpha)+\varphi(\frac{x^\alpha}{\tau_k^{\delta'}})[V_k^\beta(x^\beta)-V_k^\beta(\pm y^\alpha)])\}$$

$$\times\, e^{-iH^\alpha t_k}].$$

We will choose later $1/2 < \delta' < \delta + 1/2$. This approximation uses that the long-range interaction between the pair and the third particle mainly acts on the center of mass of the pair if the particles in the pair are sufficiently close.

<u>Lemma</u> <u>13.1</u>. With the above definitions for any $1/2 < \delta' < \delta + 1/2$

(13.5) $$\lim_{m\to\infty}\sup_{t\geq\tau_m}\|U^\alpha(t,\tau_m) - \tilde{U}^\alpha(t,\tau_m)\| = 0$$

<u>Proof</u>. The supremum in (13.5) is bounded by

(13.6) $$\sum_{k\geq m}\|\exp\{-it_k\sum_{\beta\neq\alpha}[1-\varphi(\frac{x^\alpha}{\tau_k^{\delta'}})][V_k^\beta(x^\beta)-V_k^\beta(\pm y^\alpha)]\} - \mathbb{1}\|$$

$$\leq \sum_{k\geq m} t_k\; F(|x^\alpha|\leq 2\tau_k^{\delta'})\cdot \mathrm{const}|x^\alpha|\cdot\sup_z|\nabla V_k^\beta(z)|$$

$$\leq \sum_{k\geq m} t_k\;\tau_k^{\delta'}/\tau_k^{\delta+3/2}.$$

In the first inequality we have used (7.3) and in the second (11.4). With (4.1) the last series converges and (13.5) follows. ■

We now assume that a state Ψ lies in $\mathcal{H}^{cont}(H)$ and in the orthogonal complement of the ranges of all modified wave operators.

This subspace is time-invariant and we can thus assume without loss of generality that there are $0 < E_1 < E_4 < \infty$ such that $F(H \in I(E_1,E_4))\Psi = \Psi \in \mathcal{H}^{cont}(H)$. By the observation made at the end of the last section one could restrict further to $F(E_1<H<E_4)\Psi = \Psi$. In order to show that $\Psi = 0$ we will derive asymptotic properties of Ψ which contradict each other. We denote the subspaces in question by $\mathcal{H}^s_\pm$:

$$(13.7) \qquad \mathcal{H}^{cont}(H) = \mathcal{H}^s_\pm \oplus \text{Ran } \Omega^0_{D\pm} \oplus \bigoplus_\alpha \text{Ran } \Omega^\alpha_{D\pm}.$$

First we show that $\tilde{U}^\alpha$ is a good approximate asymptotic time evolution for Ψ.

<u>Proposition</u> <u>13.2</u>. Let H satisfy (7.13)-(7.19) and (7.23). With $\tilde{U}^\alpha$ as given in (13.3), (13.4) and $F(H \in I(E_1,E_4))\Psi = \Psi \in \mathcal{H}^s_+$ there is a sequence $t_r \to \infty$ such that with a decomposition

$$(13.8) \qquad \Psi = \sum_\alpha \Psi^\alpha$$

$$(13.9) \qquad \lim_{r\to\infty} \sup_{t\geq t_r} \|\{e^{-iH(t-t_r)} - \tilde{U}^\alpha(t,t_r)\}\, e^{-iHt_r}\, \Psi^\alpha\| = 0.$$

Moreover for any given r there is a Φ^α_r such that

$$(13.10) \qquad \lim_{t\to\infty} \|e^{-iHt}\,\Psi^\alpha - \tilde{U}^\alpha(t,t_r)\,\Phi^\alpha_r\| = 0.$$

<u>Proof</u>. (13.10) is satisfied if

$$(13.11) \qquad \Phi^\alpha_r = \lim_{t\to\infty} \tilde{U}^\alpha(t,t_r)^*\, e^{-iHt}\, \Psi^\alpha$$

exists. (13.9) verifies the Cauchy criterion for this sequence.

Now fix two positive sequences $\varepsilon_r \downarrow 0$ and $E_3 \geq E_3^r \downarrow 0$ (where E_3 was determined such that it satisfies (9.10), (9.11)) and let a family of functions g^r satisfy (9.7)-(9.9) with E_3^r. Then in the corresponding decomposition (9.21) the f_i's may be chosen independent of r but the functions $f_j^{\alpha r}$ will depend on r. By Proposition 9.1 there is an N_r and arbitrary large τ_r such that with these quantities (9.30) is satisfied with $\varepsilon_r/2$. In Propositions 11.1 and 12.4 we have shown that the range of

$$(13.12) \qquad \sum_j f_j^{\alpha r}(p^\alpha/\mu^\alpha, q^\alpha/\nu^\alpha)\; f_j^{\alpha r}(x^\alpha/\tau, y^\alpha/\tau)$$

is in good approximation contained in the range of Ω_{D+}^0 for all large τ. If Q_+^s denotes the orthogonal projection onto $\mathcal{H}_+^s$ then in particular for (13.12)

$$(13.13) \qquad \lim_{\tau\to\infty} \|Q_+^s\; f_j^{\alpha r}(\cdots)\; f_j^{\alpha r}(\cdots)\| = 0.$$

Similarly by Propositions 11.4 and 12.5

$$(13.14) \qquad \lim_{\tau\to\infty} \sum_\alpha \sum_i \|Q_+^s\; f_i(q^\alpha/\nu^\alpha) f_i(y^\alpha/\tau) \sum_{j\le N_r} P_j^\alpha\| = 0.$$

In particular we get for sufficiently large τ_r with $e^{-iH\tau}\Psi = \Psi_+^s\; e^{-iH\tau}\Psi$

$$(13.15) \qquad (\mathbb{1}-Q^{N_r})e^{-iH\tau_r}\;\Psi \approx 0.$$

Combining these estimates with Proposition 9.1 there is a sequence t_r such that

$$(13.16) \qquad \|e^{-iHt_r}\;\Psi -$$

$$\sum_\alpha [g^r(h^\alpha)g^r(\mu^\alpha|x^\alpha|^2/2t_r^2) \sum_i f_i(q^\alpha/\nu^\alpha)f_i(y^\alpha/t_r)]\; e^{-iHt_r}\;\Psi\|$$

$$< \varepsilon_r.$$

We denote for such a sequence

$$(13.17) \qquad e^{-iHt_r}\,\Psi_r^\alpha := $$

$$[g^r(h^\alpha) g^r(\mu^\alpha |x^\alpha|^2/2t_r^2) \sum_i f_i(q^\alpha/\nu^\alpha)\, f_i(y^\alpha/t_r)]\, e^{-iHt_r}\,\Psi.$$

With Proposition 11.2 the future time evolution of these states is well approximated by U^α and Lemma 13.1 allows to approximate further by $\tilde{U}^\alpha$. This shows (13.9) for Ψ_r^α instead of Ψ^α. It remains to show that Ψ_r^α converge as $r \to \infty$, then (13.8) and (13.9) both hold for the limits.

By (13.16) we know that

$$(13.18) \qquad \lim_{r\to\infty} \sup_{s>r} \|\sum_\alpha (\Psi_r^\alpha - \Psi_s^\alpha)\| = 0.$$

If in addition for all α

$$(13.19) \qquad \lim_{r\to\infty} \sup_{s>r} |(\Psi_r^\alpha - \Psi_s^\alpha \;,\; \sum_{\beta\neq\alpha} (\Psi_r^\beta - \Psi_s^\beta))| = 0,$$

then the Cauchy criterion follows for each sequence Ψ_r^α separately. To show (13.19) we prove the stronger

$$(13.20) \qquad \lim_{r\to\infty} \sup_{s>r} |(\Psi_r^\alpha, \Psi_s^\beta)| = 0 \;\forall\; \alpha, \beta.$$

The following approximations are good for large r uniformly in $s > r$

$$(13.21) \qquad (\Psi_r^\alpha, \Psi_s^B)$$

$$= (e^{-iH(t_s - t_r)}\, e^{-iHt_r}\,\Psi_r^\alpha \;,\; e^{-iHt_s}\,\Psi_s^\beta)$$

$$\approx (U^\alpha(t_s, t_r) \cdots\cdots\cdots\cdots)$$

$$\approx (U^\alpha(t_s, t_r)\, e^{-iHt_r}\,\Psi_r^\alpha,\; F(|x^\beta| < v\tau_s) \cdots).$$

In the first approximation we have used Proposition 11.2, in the other v satisfies (9.10) and thus

$$\lim_{r'\to\infty} \|F(|x^\beta|>vt_s)\, g^{r'}(h^\beta)\, g^{r'}(\mu^\beta|x^\beta|^2/2t_s^2)\| = 0. \tag{13.22}$$

For any ε and all large r the absolute value of (13.21) is bounded by

$$\varepsilon + \|F(|x^\beta|<vt_s)\, U^\alpha(t_s,t_r)\, g^r(h^\alpha)\, g^r(\mu^\alpha|x^\alpha|^2/2t_r^2) \times \tag{13.23}$$

$$\times \sum_i f_i(q^\alpha/\nu^\alpha) f_i(y^\alpha/t_r)\|.$$

The estimates in the proof of Proposition 11.2 show that also the latter term is small for large r uniformly in s > r. ■

The last Proposition expresses the fact that asymptotic completeness can be violated only if the energy distribution of a two-body subsystem does not converge but is asymptotically better and better concentrated near zero. The same phenomenon must then occur for the approximate time evolution $\tilde{U}^\alpha$. In contrast to the short-range case $\tilde{U}^\alpha$ does not conserve the energy of the two body subsystems and more subtle estimates are necessary. The following form is convenient for later estimates.

Lemma 13.3. For Φ_r^α as constructed in Proposition 13.2 and for the sequence of times $\{t_s\}$ given there

$$\lim_{s\to\infty} h^\alpha\, \tilde{U}^\alpha(t_s,t_r)\, \Phi_r^\alpha = 0. \tag{13.24}$$

Proof. By construction

$$\lim_{s\to\infty} \|\tilde{U}^\alpha(t_s,t_r)\, \Phi_r^\alpha - e^{-iHt_s}\, \Psi_s^\alpha\| = 0. \tag{13.25}$$

With (13.17) and $\|h^\alpha g^s(h^\alpha)\| \to 0$ as $s \to \infty$ this implies for any $E < \infty$

$$\lim_{s\to\infty} \|F(h^\alpha<E)\ h^\alpha\ \tilde{U}^\alpha(t_s,t_r)\ \Phi_r^\alpha\| = 0. \tag{13.26}$$

It remains to show that

$$\lim_{E\to\infty} \sup_{s\geqslant r} \|F(h^\alpha>E)\ h^\alpha\ \tilde{U}^\alpha(t_s,t_r)\ \Phi_r^\alpha\| = 0. \tag{13.27}$$

This is implied by the stronger

$$\sup_{s\geqslant r} \|(h^\alpha+M)^{3/2}\ \tilde{U}^\alpha(t_s,t_r)\ \Phi_r^\alpha\| \le \text{const}, \tag{13.28}$$

where the constant M is chosen such that $h^\alpha + M \geqslant \mathbb{1}$. As a first step for that we show with τ_m,τ_n from the sequence in the definition of $\tilde{U}^\alpha$

$$\sup_{m,n}\|(h^\alpha+M)^{3/2}\ \tilde{U}^\alpha(\tau_n,\tau_m)\ (h^\alpha+M)^{-3/2}\| \le \text{const}. \tag{13.29}$$

Clearly the norm is bounded by

$$\prod_{k=m}^{n-1} \|(h^\alpha+M)^{3/2}\ U_k\ (h^\alpha+M)^{-3/2}\| \tag{13.30}$$

with the shorthand

$$U_k := \exp\{-it_k\varphi(\frac{x^\alpha}{\tau_k^\delta}) \sum_{\beta\neq\alpha} [V_k^\beta(x^\beta)-V_k^\beta(\pm y^\alpha)]\}. \tag{13.31}$$

$$\begin{aligned} (13.32)\quad & \|(h^\alpha+M)^{3/2}\ U_k\ (h^\alpha+M)^{-3/2}\| \\ & \le \|(h^\alpha+M)^{1/2}\ [h_0^\alpha,U_k]\ (h^\alpha+M)^{-3/2}\| + \|(h^\alpha+M)^{1/2}\ U_k\ (h^\alpha+M)^{-1/2}\| \\ & \le \text{const}\{\|[h_0^\alpha,U_k]\ (h^\alpha+M)^{-3/2}\| + \sup_j\|p_j\ [h_0^\alpha,U_k]\ (h^\alpha+M)^{-3/2}\|\} \\ & + \{1 + \|(h^\alpha+M)^{-1/2}\ U_k^*\ [h_0^\alpha,U_k]\ (h^\alpha+M)^{-1/2}\|\}^{1/2} \\ & \le 1 + c_k. \end{aligned}$$

The derivative of the exponent in (13.31) is bounded by

$$\text{const } t_k/(\tau_k)^{\delta+3/2}$$

and higher derivatives decay even faster in k. (c.f. (13.41) and (13.42) below.) Moreover

$$\max_j \|p_j(h^\alpha+M)^{-1/2}\| + \max_{i,j} \|p_i\, p_j(h^\alpha+M)^{-1}\| \le \text{const.}$$

Then it is easy to estimate the commutator terms and obtain

$$c_k \le \text{const } t_k/\tau_k^{\delta+3/2}. \tag{13.33}$$

This is summable and therefore arbitrary products (13.30) are uniformly bounded. The same estimates apply for "partial" intervals (13.3) and consequently

$$\sup_{t,T\geq 1} \|(h^\alpha+M)^{3/2}\ \tilde{U}^\alpha(t,T)\ (h^\alpha+M)^{-3/2}\| \le \text{const.} \tag{13.34}$$

The same bound holds for $(\tilde{U}^\alpha)^*$ since we have not used the sign of the exponent in (13.31).

Now (13.28) follows from (13.34) if $\Phi_r^\alpha \in \mathfrak{D}((h^\alpha+M)^{3/2})$. The latter follows since for the approximating sequence of Φ_r^α

$$\|(h^\alpha+M)^{3/2}\ [\tilde{U}^\alpha(t_s,t_r)]^*\ e^{-iHt_s}\ \Psi_s^\alpha\| \tag{13.35}$$

$$\le \sup_s \|(h^\alpha+M)^{3/2}\ [\tilde{U}^\alpha(t_s,t_r)]^*\ (h^\alpha+M)^{-3/2}\|\cdot\|(h^\alpha+M)^{3/2}\ g^s(h^\alpha)\|$$

is uniformly bounded in s. ■

Next we derive an upper bound for the energy of the pair at late times. It will be used below to show that asymptotically the internal motion of the pair is trivial.

Lemma 13.4. Let for the (negative part of the) long-range potential (7.18) be satisfied with $\delta > \sqrt{3} - 3/2$. With τ_k as given in the definition of $\tilde{U}^\alpha$ there is a $\delta'' > 1$ such that

$$(13.36) \qquad \| h^\alpha \, \tilde{U}^\alpha(t,t_r) \, \Phi_r^\alpha \| \le \mathrm{const}(\tau_k)^{-\delta''}$$

for $\tau_{k+1} \geqslant t > \tau_k$.

Proof. We set

$$(13.37) \qquad a(t) := \| h^\alpha \, \tilde{U}^\alpha(t,t_r) \, \Phi_r^\alpha \| .$$

$$(13.38) \qquad |a(t) - a(\tau_k)| \le$$

$$\| [h^\alpha, \exp\{-i(t-\tau_k) \sum_{\beta\neq\alpha} (\varphi(\frac{x^\alpha}{\tau_k^\delta}) [V_k^\beta(x^\beta) - V_k^\beta(\pm y^\alpha)])\}] \times$$

$$\times \, e^{-iH^\alpha(t-\tau_k)} \, \tilde{U}^\alpha(\tau_k, t_r) \, \Phi_r^\alpha \| .$$

$$(13.39) \quad 2\mu^\alpha \| [h^\alpha, \exp\{\cdots\}] \, e^{-iH^\alpha(t-\tau_k)} \, \tilde{U}^\alpha(\tau_k, t_r) \, \Phi_r^\alpha \|$$

$$\le 2\cdot\nu\cdot\max_j (t-\tau_k) \sum_{\beta\neq\alpha} \| \frac{\partial}{\partial x_j^\alpha} (\varphi(\frac{x^\alpha}{\tau_k^\delta}) [V_k^\beta(\mathrm{const}\; x^\alpha \pm y^\alpha) - V_k^\beta(\pm y^\alpha)]) \, p_j^\alpha \times$$

$$\times \exp\{\cdots\} e^{-iH^\alpha(t-\tau_k)} \, \tilde{U}^\alpha(\tau_k, t_r) \Phi_r^\alpha \|$$

$$+ \, (t-\tau_k)^2 \| \sum_{\beta\neq\alpha} \frac{\partial}{\partial x^\alpha} (\varphi[\cdots]) \|^2 + (t-\tau_k) \| \sum_{\beta\neq\alpha} \Delta(\varphi[\;]) \| .$$

We use that

$$(13.40) \qquad \mathrm{supp}(\nabla\varphi)(\frac{x^\alpha}{\tau_k^\delta}) \subset \{x^\alpha \,|\, \tau_k^{\delta'} \le |x^\alpha| \le 2\tau_k^{\delta'}\}$$

and obtain by (7.19) the estimates for the gradient

$$(13.41) \qquad \| \sum_{\beta\neq\alpha} \frac{\partial}{\partial x^{\alpha}}(\varphi[\cdots])\| \le \mathrm{const}(\tau_k)^{-\delta-3/2},$$

and for second derivatives w.r.t. x^{α} with $\delta' > 1/2$

$$(13.42) \qquad \| \sum_{\beta\neq\alpha} D^{\gamma}(\varphi[\ ])\| \le \mathrm{const}(\tau_k)^{-\delta-2}.$$

Thus the last two terms in (13.39) are bounded uniformly in $\tau_k < t \le \tau_k + t_k = \tau_{k+1}$ by

$$(13.43) \qquad \mathrm{const}(t-\tau_k)/(\tau_k)^{\delta+2} \le \mathrm{const}(t-\tau_k)/(\tau_{k+1})^{\delta+2}.$$

For the first term on the right hand side of (13.39) we take the square of the norm

$$(13.44) \quad b_j^2 := \|\nabla_j(\varphi[\cdots])\ p_j^{\alpha} \cdots\|^2$$

$$= 2i(\exp\{\ \}\ e^{-iH^{\alpha}(t-\tau_k)}\ \tilde{U}^{\alpha}(\tau_k,t_r)\ \Phi_r^{\alpha}\ ,$$

$$(\nabla_j)^2(\varphi[\cdots])\cdot\nabla_j(\varphi[\cdots])\ p_j^{\alpha}\ \exp\{\ \}\ e^{-iH^{\alpha}(t-\tau_k)}\ \tilde{U}^{\alpha}(\tau_k,t_r)\ \Phi_r^{\alpha})$$

$$+ (\cdots\ \Phi_r^{\alpha}\ ,\ |\nabla_j(\varphi[\cdots])|^2\ (p_j^{\alpha})^2\ \cdots\ \Phi_r^{\alpha}).$$

Then we get the estimate

$$(13.45) \quad \sum_j b_j^2 \le \mathrm{const}(\tau_k)^{-\delta-2}\ b_j + \mathrm{const}(\tau_k)^{-2\delta-3}\ a(\tau_{k+1})$$

$$+ \sum_j \frac{1}{2\mu^{\alpha}}\ \|\,|\nabla_j(\varphi[\cdots])|^2\ V_-^{\alpha}(x^{\alpha})\ (h^{\alpha}-i)^{-1}\|\cdot\|(h^{\alpha}-i)\cdots\|.$$

In the last term the second factor is uniformly bounded and the first norm is bounded by

$$(13.46)\qquad \text{const}(\tau_k)^{-3-2\delta}\ \|F(|x^\alpha|>\tau^{\delta'})\ V_-^\alpha(x^\alpha)\ (h^\alpha-i)^{-1}\|$$

$$\le \text{const}(\tau_k)^{-3-2\tau-\delta'(\delta+1/2)}.$$

If

$$(13.47)\qquad \delta + 1/2 > \delta' > (\sqrt{3} - 1) < 3/4$$

then the exponent in (13.46) is $-2(1 + \delta'')$ where

$$(13.48)\qquad 1 < \delta'' = (\delta+1/2)\ (1+\delta'/2) < 1 + \delta.$$

Now the bound (13.45) implies

$$(13.49)\qquad b_j \le c'\ (\tau_k)^{-1-\delta''} + d\ (\tau_k)^{-\delta-3/2}\ \sqrt{a(\tau_{k+1})}.$$

We insert it into (13.39), (13.38) to get

$$(13.50)\qquad |a(\tau_{k+1})-a(\tau_k)|$$

$$\le t_k\ \{c\ (\tau_{k+1})^{-1-\delta''} + d\ \sqrt{a\ (\tau_{k+1})}(\tau_{k+1})^{-\delta-3/2}\}$$

for some constants c and d. Moreover at intermediate times

$$(13.51)\qquad \sup_{\tau_k\le t\le\tau_{k+1}} |a(t)-a(\tau_k)| \to 0 \text{ as } k \to \infty.$$

The asymptotic property (13.24) implies then

$$(13.52)\qquad \liminf_{k\to\infty} a(\tau_k) = 0.$$

Since $a(\tau_k)$ is uniformly bounded one can sum up the inequality (13.50) once to obtains

$$(13.53)\qquad a(\tau_n) \le \sum_{k=n}^{\infty} \text{const}\ t_k\ (\tau_{k+1})^{-\delta-3/2} + \liminf_{k\to\infty} a(\tau_k)$$

$$\le \text{const}(\tau_n)^{-\delta-1/2}.$$

Since

$$(13.54)\qquad \delta + 3/2 + \tfrac{1}{2}(\delta+\tfrac{1}{2}) < 1 + \delta''$$

it is sufficient to insert (13.53) into (13.50) and obtain

$$(13.55)\qquad a(\tau_n) \le \text{const} \sum_{k=n}^{\infty} t_k\ (\tau_k)^{-1-\delta''} \le \text{const}(\tau_k)^{-\delta''}.$$

This is the desired bound (13.36). ■

It is only here that we use the restriction $\delta > \sqrt{3} - 3/2$ rather than $\delta > 0$. We use it to obtain the bound (13.36) with some δ'' > 1. If only $\delta > 0$ we would still get some lower bound for δ''. We need the restriction of δ to have efficient control of p_j by $|h^\alpha|^{1/2}$. The error has suitable decay if $|x^\alpha|$ is large enough. The latter is guaranteed by the lower cutoff φ in the exponents of $\tilde{U}^\alpha$. The lower bound on δ' used there pushes δ up as well.

Lemma 13.5. Let $\Psi \in \mathcal{H}_+^s$ be as above and $\delta > \sqrt{3} - 3/2$. Then

$$(13.56)\qquad \lim_{n\to\infty} \sup_{t\geq\tau_n} \|\{e^{-iH(t-\tau_n)} - \bar{U}^\alpha(t,\tau_n)\}\ e^{-iH\tau_n}\ \Psi\| = 0$$

where

$$(13.57)\qquad \bar{U}^\alpha(\tau_n,\tau_m) = \prod_{k=m}^{n-1}{}' \exp\{-it_k V_k^\beta(\pm y^\alpha)\}\ e^{-ik_0^\alpha t_k}$$

and analogously for intermediate times.

Proof. By the above it is sufficient to show that for all α

$$(13.58) \qquad \lim_{n\to\infty} \sup_{t\geq\tau_n} \| \{\tilde{U}^\alpha(t,\tau_n) - \bar{U}^\alpha(t,\tau_n)\} \tilde{U}^\alpha(\tau_n, t_r) \Phi_r^\alpha \| = 0.$$

As a first step we drop the $t_k h^\alpha$ from $\tilde{U}^\alpha$:

$$(13.59) \qquad \bar{U}^\alpha(\tau_n, \tau_m) := \prod_{k=m}^{n-1}{}' \exp\{\cdots\} e^{-ik_0^\alpha t_k}$$

with the same exponent as in the definition of $\tilde{U}^\alpha$. Then

$$(13.60) \quad \sup_{t\geq\tau_k} \| \{\tilde{U}^\alpha(t,\tau_k) - \bar{U}^\alpha(t,\tau_k)\} \tilde{U}^\alpha(\tau_k, t_r) \Phi_r^\alpha \|$$

$$\leq \sum_{n\geq k} t_n \| h^\alpha \tilde{U}^\alpha(\tau_n, t_r) \Phi_r^\alpha \| \leq \text{const} \sum_{n\geq k} t_n/(\tau_n)^{\delta''} < \varepsilon/2$$

for some large k by Lemma 13.4. For this k choose some $R = R(\varepsilon)$ such that

$$(13.61) \qquad \| F(|x^\alpha| > R) \tilde{U}^\alpha(\tau_k, t_r) \Phi_r^\alpha \| < \varepsilon/4.$$

Then

$$(13.62) \quad \sup_{t\geq\tau_n} \| \{\tilde{U}^\alpha(t,\tau_n) - \bar{U}^\alpha(t,\tau_n)\} \tilde{U}^\alpha(\tau_n, t_r) \Phi_r^\alpha \|$$

$$\leq \varepsilon + \sup_{t\geq\tau_n} \| \{\bar{U}^\alpha(t,\tau_n) - \bar{U}^\alpha(t,\tau_n)\} \bar{U}^\alpha(\tau_n, \tau_k) F(|x^\alpha| < R) \|.$$

But the norm on the right hand side of (13.62) is zero if n is large enough such that $\tau_n^{\delta'} > R$ by the support properties of $\varphi(|x^\alpha|/\tau_k^{\delta'})$. ■

Note that under $\bar{U}^\alpha$ there is no internal motion of the pair. In particular there are for any ε an E_3 and N such that

$$\sum_\alpha \sup_t \| \{F(h^\alpha < E_3) P^{cont}(h^\alpha) + \sum_{j>N} P_j^\alpha\} e^{-iHt} \Psi \| < \varepsilon.$$

Thus condition (12.17) is satisfied and by Proposition 12.6 Ψ lies in the direct sum of the ranges of the modified wave operators. Our assumption in this section that Ψ is orthogonal to the ranges of all wave operators therefore implies that $\Psi = 0$. We have shown:

Theorem 13.6. Let H satisfy (7.13)-(7.19) with $\delta > \sqrt{3} - 3/2$ and assume that (7.23) holds. Then the modified Dollard wave operators (12.29)-(12.31) exist and are complete. In particular $H \upharpoonright \mathcal{H}^{cont}(H)$ is unitarily equivalent to a direct sum of free Hamiltonians (Laplacian operators) and it has no singular continuous spectrum.

XIV. Concluding Remarks and Notes

There are several points where the results presented here can be improved. Only in the last section we have imposed the stronger condition $\delta > \sqrt{3} - 3/2$. With that condition we had then derived that the asymptotic internal motion of the pair would have to be trivial. However, for a much wider class of internal motions of the pair the particular behaviour with decreasing energy and increasing separation should lead to the desired contradiction. It should not be hard to eliminate the positive lower bound of δ and extend this completeness proof to all $\delta > 0$. This is a purely mathematical question since the only long-range potential of physical interest--the Coulomb potential--has $\delta = 1/2$ and it is covered by the results given here.

Another restriction of the proof given here is the decay assumption (7.23) for any zero energy bound states of the two-body subsystems. It is disturbing because it is an implicit condition on the potentials. We have shown in [13] with a slightly different argument that asymptotic completeness holds without that condition if only short range potentials are present. Most of that proof can be extended to the long-range case. There is a remaining problem in this approach. We do not know whether U_b^{α} is a good asymptotic time evolution if the pair is in a bound state with slow decay, in particular in low dimensions where the spreading of the wave functions is not sufficient to guarantee convergence of the Cook integral. We believe that this difficulty can be overcome but the full proof would then still be special for three-body systems. Large parts of the proof given here, however, carry over with suitable generalizations to the N-body case. Therefore it is an interesting open question whether the results, in particular those of Section VIII on asymptotic observables, can be shown without the decay property (7.23).

There are several directions of further study. One can weaken the conditions on local singularities and admit form bounded and highly singular positive potentials. This does not seem to be a rewarding study and we do not expect new insights from it. Many quantities have to be regularized and numerous purely technical estimates would have to be carried out. The inclusion of velocity dependent forces (pseudodifferential operators as potentials) and of three-body forces does not cause any problems, we have omitted them here only for simplicity of presentation.

A more interesting generalization is the inclusion of potentials with very long range where our simple minded intermediate time evolution and the asymptotic Dollard time evolution are not applicable. There are many technically different possibilities for two-body systems. The method closest in spirit to the presentation here is the comprehensive treatment of H. Kitada and K. Yajima [27]. They use Fourier integral operators, i.e. phase space localization plays a major role. The extension to three-body systems is presently being carried out by M. Combescure [4].

The main challenge is to treat systems with arbitrary particle number. Many results are known for special classes of potentials. Most of them are summarized in the book of I. M. Sigal [48], see its introduction or [25] as a guide to the literature. For other results see [21] and for two cluster scattering [8] and references given there. The so called "generalized three-body systems" have in common that the only scattering channels are two-cluster channels and the totally free one. If for an N-body system there are no bound states with less then N-1 particles, then the system belongs to this class. E. Mourre and I. M. Sigal [36] and M. Krishna [30] recently gave geometrical completeness proofs for this class in the short-range case.

Another possible generalization is to study more general operators than Schrödinger operators, i.e. to replace the Laplacian by other "free Hamiltonians" like Klein-Gordon- or Dirac- operators. An extremely large class of operators can be treated in the two-body short-range case, but the class seems to shrink when long-range potentials or higher particle numbers are considered, see e.g. [39], [51].

In these notes we have restricted our attention to scattering theory. The absence of a singular continuous spectrum follows automatically and with little extra work one could show that bound state energies of the tree-body system (counting multiplicity) can accumulate only at thresholds (two-body bound state energies). These questions can be treated directly without studying the existence and completeness of scattering. The strongest results for N-body systems are due to Perry, Sigal, and Simon [44] (see there for references to earlier work) with simplifications of Froese and Herbst [17]. They are based on Mourre's work for 3-body systems [33].

General references to papers on two-body scattering theory can be found e.g. in the notes of [46], [2], [3], [23], [41]. The systematic study of the geometrical time-dependent scattering theory started with [6] and various modifications, simplifications, improvements, and extensions appeared within the next few years. For references see the book of P. Perry [43] or the review article [9]. The parallel development of "conjugate operator"-methods of Mourre [32,34] has common deep roots with the present approach. Related ideas to phase space analysis had been used in other areas like e.g. partial differential equations (microlocal analysis) or constructive quantum field theory. The use of asymptotic observables to control propagation in phase space was introduced in [9] and later extended in [12] and by

Sinha and Muthuramalingam [49,40]. The intermediate time evolution U for long range potentials was announced in [11] and is published here for the first time, extensions will appear in [15].

The first proof of asymptotic completeness for certain three-body quantum systems was given by L. D. Faddeev [16]. Many improvements were made by several authors in the following years. This development up to 1976 is summed up e.g. in the review of J. Ginibre [19]. The basic restrictions of that approach were (i) space dimension $\nu \geq 3$, (ii) pair potentials decay faster than the second inverse power of the distance, (iii) absence of zero-energy resonances or bound states for two body subsystems. For more recent related results see [20] and references in [48]. The essentials of the first geometrical, time-dependent proof of three-body completeness were given in [10], more details and some improvements appeared in [13]. The above restrictions (i) and (iii) were eliminated completely and the decay requirement (ii) was reduced by one power to cover the general short-range case. Simultaneously E. Mourre extended his method of conjugate operators to cover three-body systems as well [35]. See also Sinha, Krishna, Muthuramalingam [50]. With generalization to higher particle numbers in view the results on asymptotic observables were strengthened in [14], additional improvements led to the presentation given here in Sections VIII and IX.

The three-body completeness problem including Coulomb forces was first treated by S. P. Merkuriev [31]. The proof is based on a modification of Faddeev's equations, it is very complicated and depends sensitively on the Coulombic tail of the potentials. However, it has the advantage that it rigorously establishes equations which can be evaluated numerically.

These lectures give the first proof of asymptotic completeness for three-body systems when the potentials belong to a large class of long-range potentials and general short-range potentials. A drawback is the use of the weak implicit decay assumption (7.23).

Acknowledgements. I am very grateful to Sandro Graffi for organizing the extremely stimulating Session on Schrödinger operators and to Professor Conti for the generous arrangements of CIME.

A large part of these lecture notes was written during my stay at the California Institute of Technology. I am indebted to many colleagues for helpful discussions and to the S. Fairchild Foundation for financial support.

References

[1] **P. K. Alsholm, T. Kato**: Scattering with long-range potentials, in: *Partial Differential Equations*, Proc. Symp. Pure Math. 23, Amer. Math. Soc. 1973, pp. 393-399.

[2] **W. O. Amrein, K. M. Jauch, K. B. Sinha**: *Scattering Theory in Quantum Mechanics*, Benjamin, Reading 1977.

[3] **H. Baumgärtel, M. Wollenberg**: *Mathematical Scattering Theory*, Akademie Verlag, Berlin and Birkhäuser, Basel 1983.

[4] **M. Combescure**: Propagation and local decay properties for long-range scattering of quantum three-body systems, preprint LPTHE Orsay 84/6, 1984.

[5] **J. Dollard**: Asymptotic convergence and the Coulomb interaction, J. Math. Phys. 5, 729-738(1964); Quantum mechanical scattering theory for short-range and Coulomb interactions, Rocky Mt. J. Math. 1, 5-88(1971).

[6] **V. Enss**: Asymptotic completeness for quantum mechanical potential scattering, I. Short range potentials, Commun. Math. Phys. 61, 285-291(1978).

[7] ——————: ———, II. Singular and long-range potentials, Ann. Phys. 119, 117-132(1979); and addendum, preprint Bielefeld BI-TP79/26, 1979, unpublished.

[8] —————— Two cluster scattering of N charged particles, Commun. Math. Phys. 65, 151-165(1979).

[9] —————— Geometric methods in spectral and scattering theory of Schrödinger operators, in: *Rigorous Atomic and Molecular Physics*; G. Velo and A. S. Wightman eds., Plenum, New York 1981, pp. 1-69 (Proceedings Erice 1980).

[10] —————— Completeness of three body quantum scattering, in: *Dynamics and Processes*, P. Blanchard, L. Streit eds., Springer Lecture Notes in Mathem. 1031, Berlin 1983, pp. 62-88 (Proceedings Bielefeld 1981).

[11] —————— Propagation properties of quantum scattering states, J. Func. Anal. 52, 219-251(1983).

[12] —————— Asymptotic observables on scattering states, Commun. Math. Phys. 89, 245-268(1983).

[13] —————— Scattering and spectral theory for three-particle systems, in: *Differential Equations*, I. W.

Knowles, R. T. Lewis eds., North Holland Mathematics Studies vol. 92, Amsterdam 1984, pp. 173-204 (Proceedings Birmingham 1983).

[14] ———————— Topics in scattering theory for multiparticle quantum mechanics, a progress report, Physica 124A, 269-292(1984)(Proceedings Boulder 1984).

[15] **V. Enss, M. Knick, M. Schneider**: Approximate quantum time-evolutions for long-range potentials, in preparation.

[16] **L. D. Faddeev**: *Mathematical Aspects of the Three Body Problem in the Quantum Scattering Theory*, Israel Program for Scientific Translations, Jerusalem 1965.

[17] **R. Froese, I. Herbst**: A new proof of the Mourre estimate, Duke Math. J. 49, 1075-1085(1982).

[18] **R. Froese, I. Herbst, M. and T. Hoffmann-Ostenhoff**: On the absence of positive eigenvalues for one-body Schrödinger operators, J. Anal. Math. 41, 272-284(1982).

[19] **J. Ginibre**: Spectral and scattering theory of the Schrödinger equation for three-body systems, in: *The Schrödinger Equation*, W. Thirring, P. Urban eds., Acta Phys. Austr., Suppl. 17, 95-138(1977)(Proceedings Vienna 1976).

[20] **G. A. Hagedorn, P. A. Perry**: Asymptotic completeness for certain three-body Schrödinger operators, Commun. Pure Appl. Math. 36, 213-232(1983).

[21] ————————: Completeness in four-body scattering, preprint 1983.

[22] **L. Hörmander**: The existence of wave operators in scattering theory, Math. Z. 146, 69-91(1976).

[23] ————————: *The Analysis of Linear Partial Differential Operators* I-IV , Springer, Berlin 1983 - 1985.

[24] **W. Hunziker**: On the space-time behavior of Schrödinger wavefunctions, J. Math. Phys. 7, 300-304(1966).

[25] **W. Hunziker, I. M. Sigal**: Review on N-body scattering theory, in preparation.

[26] **A. Jensen, E. Mourre, P. A. Perry**: Multiple commutator estimates and resolvent smoothness in quantum scattering theory, Ann. Inst. H. Poincaré A41, 207-225(1984).

[27] **H. Kitada, K. Yajima**: A scattering theory for time-dependent long-range potentials, Duke Math. J. 49, 341-376 (1982).

[28] **M. Knick**: Diplomarbeit, in preparation.

[29] **M. Krishna**: Time decay properties of N-body quantum states with low energy, Indian Statistical Institute Techn. Rep. 8412, 1984.

[30] ———————— Large time behaviour of some N-body systems, Commun. Math. Phys. to appear.

[31] **S. P. Merkuriev**: On the three-body Coulomb scattering problem, Ann. Phys. 130, 395-426(1980), and references given there.

[32] **E. Mourre**: Link between the geometrical and the spectral transformation approaches in scattering theory, Commun. Math. Phys. 68, 91-94(1979).

[33] ———————— Absence of singular spectrum for certain self-adjoint operators, Commun. Math. Phys. 78, 391-408(1981).

[34] ———————— Operateurs conjugués et propriétés de propagation, Commun. Math. Phys. 91, 279-300(1983).

[35] ———————— Operateurs conjugués et propriétés de propagation II, remarques sur la complétude asymptotique des systemes a trois corps, preprint CPT82/P. 1379, CNRS Marseille 1982.

[36] **E. Mourre, I. M. Sigal**: Phase-space analysis and scattering theory for N-particle systems (Micro-local analysis of propagation of particles), preprint 1984.

[37] **PL. Muthuramalingam**: Asymptotic completeness for the Coulomb potential, Indian Statistical Inst. Techn. Rep., 1982.

[38] ———————— Lectures on spectral properties of the (two body) Schrödinger operator $-1/2\,\Delta + W(Q)$ on $L^2(\mathbb{R}^n)$ using time dependent scattering theory in quantum mechanics, Indian Statistical Inst., preprint 1983.

[39] ———————— Spectral properties of vaguely elliptic pseudo-differential operators with momentum-dependent long-range potentials using time-dependent scattering

theory, J. Math. Phys. 25, 1881-1899(1984);
A note on time dependent scattering theory for $P_1^2-P_2^2+(1+|Q|)^{-1-\varepsilon}$ and $P_1P_2+(1+|Q|)^{-1-\varepsilon}$ on $L^2(\mathbb{R}^2)$, Math. Z. to appear; and further preprints.

[40] **PL. Muthuramalingam and K. B. Sinha**: Asymptotic completeness in long range scattering II, Ann. scient. Ec. Norm. Sup. 18, 57-87 (1985).

[41] **R. G. Newton**: *Scattering Theory of Waves and Particles*, second ed., Springer, New York 1982.

[42] **P. A. Perry**: Propagation of states in dilation analytic potentials and asymptotic completeness, Commun. Math. Phys. 81, 243-259(1981).

[43] ——————— *Scattering Theory by the Enss Method*, Harwood, Chur, 1983 (Mathematical Reports 1, part 1).

[44] **P. A. Perry, I. M. Sigal, B. Simon**: Spectral analysis of N-body Schrödinger operators, Ann. Math. 114, 519-567(1981).

[45] **C. Radin, B. Simon**: Invariant domains for the time-dependent Schrödinger equation, J. Diff. Equ. 29, 289-296(1978).

[46] **M. Reed, B. Simon**: Methods of Modern Mathematical Physics, III. Scattering Theory, Academic Press, New York 1979.

[47] **M. Schneider**: Asymptotische Lösung der Schrödinger-Gleichung für langreichweitige Potentiale, Diplomarbeit, Bochum 1984.

[48] **I. M. Sigal**: *Scattering Theory for Many-Body Quantum Mechanical Systems-Rigorous Results*, Springer Lecture Notes in Math. 1011, Berlin 1983.

[49] **K. B. Sinha, PL. Muthuramalingam**: Asymptotic evolution of certain observables and completeness in Coulomb Scattering I, J. Func. Anal. 55, 323-343(1984).

[50] **K. B. Sinha, M. Krishna, PL. Muthuramalingam**: On the completeness in three body scattering, Ann. Inst. H. Poincaré, A41, 79-101(1984).

[51] **B. Thaller, V. Enss**: Asymptotic Observables and Coulomb Scattering for the Dirac Equation, preprint Berlin Nr. 198, 1985.

Some Aspects of the Theory of Schrödinger Operators *

Barry Simon
Division of Physics, Mathematics and Astronomy
California Institute of Technology
Pasadena, California 91125

In these notes, we will survey a part of theory of the operator $-\Delta + V$. More extensive surveys can be found in [1,2,3] and in [4].

1. Self-adjointness, properties of eigenfunctions and all that

There is an enormous literature on the basic issue of giving a domain where $-\Delta + V$ is self-adjoint or essentially self-adjoint. To a large extent, I think one can single out two results as the most important: (1) The basic perturbation results of Kato-Rellich which accomodate virtually all cases of physical interest (2) "Kato's inequality," which, at least among positive V, is definitive. We will describe the first result briefly (for background on definition of self-adjoint, etc., see [5,6,7]; for a discussion of Kato's inequality, see [1,8,9,10]).

Theorem 1.1 (The Kato-Rellich theorem [11,12]) Let A be a self-adjoint operator on a Hilbert space, $\mathcal{H}$, and let B be symmetric. Suppose that $D(B) \supset D(A)$ and for some $a < 1$ and $b < \infty$,

$$\|B\varphi\| \leq a\|A\varphi\| + b\|\varphi\| \tag{1.1}$$

for all $\varphi \in D(A)$. Then $A+B$ is self-adjoint on $D(A)$ and any core for A is a core for $A+B$.

For a proof, see [1], pp. 162-163.

To apply this to $-\Delta + V$, we set $A = -\Delta$, $B = V$ and study (1.1). In this form, (1.1) is related to Sobolev estimates. Kato studied when (1.1) held in terms of L^p-spaces a point of view I long preferred, but I have come around to prefer a point of view introduced by Stummel [13].

Definition Fix $\nu \geq 4$, and $0 < \alpha < 4$ and let $S_\alpha^{(\nu)}$ be the set of functions, V, on R^ν obeying

---

* Research partially supported by USNSF grant MCS-81-20833

$$\sup_x \int_{|y-x|\leq 1} |x-y|^{-(\nu-4+\alpha)} |V(y)|^2 dy < \infty \tag{1.2}$$

If $\nu \leq 3$, we define $S_\alpha^{(\nu)}$ in terms of (1.2) with $|x-y|^{-(\nu-4+\alpha)}$ replaced by 1 (independently of α).

With these definitions, it is not hard to prove the following pair of results (see Stummel [13]).

Theorem 1.2 If $V \in S_\alpha^{(\nu)}$, then (1.1) holds on $D(-\Delta)$ where $B = V$, $A = -\Delta$ and a can be taken arbitrarily close to zero.

Theorem 1.3 If $g \in S_\alpha^{(\mu)}$, $\nu > \mu$ and $V(x) = g(\pi x)$ where π is a linear map of R^ν *onto* R^μ, then $V \in S_\alpha^{(\nu)}$.

Thm. 1.2 is proven by noting that for $\nu \geq 4$, the integral of $(-\Delta+\kappa^2)^{-2}$ goes to $|x-y|^{-(\nu-4)}$ for $|x-y|$ small and as $e^{-\kappa|x-y|}$ for $|x-y|$ large and for $\nu \leq 3$, the kernel is bounded at small distances. As a result, $\|V(-\Delta+\kappa^2)^{-2}V\| \to 0$ as $\kappa \to \infty$. Theorem 1.3 follows by noting that $|x-y|^{-(\nu-4+\alpha)}$ integrated over $\nu-\mu$ variables (and cutoff at large distances) is bounded by $|x-y|^{-(\mu-4+\alpha)}$.

The most important special case of Thm. 1.3 is to take μ fixed ($\mu=3$ is the physical case), $\nu = \mu N$, write a point in R^ν as $x = (x_1,\ldots,x_N)$ with $x_j \in R^\mu$ and let $Tx = x_i - x_j$. Thus picking, for all pairs i,j, a function $\upsilon_{ij} \in S_\alpha^{(\mu)}$ and letting $V_{ij}(x) = \upsilon_{ij}(x_i - x_j)$, we see that $V_{ij} \in S_\alpha^{(\nu)}$. Therefore, the operator

$$\tilde{H} = \tilde{H}_0 + V; \quad \tilde{H}_0 = \sum_{j=1}^{N} -(2m_j)^{-1}\Delta_j; \quad V = \sum_{i<j} V_{ij} \tag{1.3}$$

called an *N-body Hamiltonian* obeys

Theorem 1.4 Any N-body Hamiltonian with $\upsilon_{ij} \in S_\alpha^{(\mu)}$ defines an operator $\tilde{H}$ self-adjoint on $D(-\Delta)$ and essentially self-adjoint on $C_0^\infty(R^\nu)$.

* * *

We used $\tilde{H}$ for the operator in (1.3) because there is a closely related operator, H, on $L^2(R^{\mu(N-1)})$ called the *operator with center of mass removed*. Here are two ways of understanding this change:

(1) Let $R = \sum m_i x_i / \sum m_i$ and let $\zeta_1, \ldots, \zeta_{N-1}$ be N - 1 additional μ-component coordinates (i.e. linear functions of the x's), so that (i) ζ_j is invariant under $x_j \to x_j + a$ for any a (ii) $x_j \to R, \zeta_j$ is an invertible transformation. For example, one might take

$$\zeta_j = r_j - r_N \qquad j=1,\ldots,N-1 \tag{1.4}$$

Then by writing $R^{\mu N} = R^{\mu} \times R^{\mu(N-1)}$ by the coordinates $R, \zeta, L^2(R^{\mu N})$ decomposes into $\mathcal{H}_{cm} \otimes \mathcal{H} \equiv L^2(R^{\mu}) \otimes L^2(R^{\mu(N-1)})$ (functions of R tensored by functions of ζ). Under this decomposition

$$\tilde{H} = H_{0,cm} \otimes 1 + 1 \otimes H \tag{1.5}$$

where $H_{0,cm} = -2(\sum m_i)^{-1} \Delta_R$ and $H = H_0 + V$. The precise form of H_0 depends on the choice of local coordinates. For example, in the coordinate system (1.4),

$$H_0 = -\sum_{j=1}^{N-1} (2\mu_j)^{-1} \Delta_{\zeta_j} + m_N^{-1} \sum_{i<j} \nabla_i \cdot \nabla_j \tag{1.6}$$

with $\mu_j = (m_N^{-1} + m_j^{-1})^{-1}$.

(2) ([14,15]) View $\tilde{H}_0$ as one half the Laplace Beltrami operator associated to the metric $\|dx\|^2 = \sum m_i (dx_i)^2$. Let $X = \{x \mid \sum m_i x_i = 0\}$. Then in the metric, $X^{\perp} = \{x \mid x_1 = x_2 = \cdots = x_N\}, \mathcal{H}_{cm} = L^2(X^{\perp}), \mathcal{H} = L^2(X)$ and H_0 is just the Laplace-Beltrami operator on X in the induced metric.

For later purposes, we introduce some additional notation to describe N-body systems. A partition of $\{1,\ldots,N\}$, i.e. a family $C_1,\ldots,C_k$ of disjoint subsets of $\{1,\ldots,N\}$ which exhaust $\{1,\ldots,N\}$ is called a *cluster decomposition*. We write $a = \{C_1,\ldots,C_k\}$; $k \equiv \#(a)$. The family of cluster decompositions is important because in various aspects of the study of N-body Hamiltonians, one expects that we want to analyze what happens as $|x| \to \infty$ with $\sum m_i x_i = 0$. This happens if the system breaks up into distinct clusters; i.e. we can find numbers $R_1,\ldots,R_k$ and a decomposition a so $|x_i - R_j|$ stays bounded if $i \in C_j$ and so each $|R_i - R_j|$ goes to infinity.

Given a, we pick coordinates $\zeta_1,\ldots,\zeta_k$ involving differences of center of mass of clusters in a, and "internal coordinates," $\zeta^1,\ldots,\zeta^{N-1-k}$, i.e. coordinates left invariant by the transformations $x_i \to x_i + R_{j(i)}$ where $j(i)$ is that j with $x_i \in C_j$. (Put differently, $\zeta^1,\ldots,\zeta^{N-1-k}$ are coordinates for the plane $X^a = \{x \mid \sum_{i\in C_j} m_i x_i = 0, \text{ all } j\}$ and $\zeta_1,\ldots,\zeta_k$ for its orthogonal complement, X_a, in X). If we decompose $\mathcal{H} = \mathcal{H}^a \otimes \mathcal{H}_a$ corresponding to functions of ζ^a and ζ_a (i.e. $\mathcal{H}^a = L^2(X^a), \mathcal{H}_a = L^2(X_a)$), then we have

$$V = V(a) + I(a);\ I(a) = \sum_{(ij)\not\subset a} V_{ij};\ V(a) = \sum_{(ij)\subset a} V_{ij}$$

(where $(ij)\subset a$ means i and j are in the same cluster), and

$$H \equiv H(a) + I(a)$$

$$H(a) = H^a \otimes I + I \otimes T_a$$

where T_a has no potentials and is exactly the kinetic energy of relative motion of the clusters. Eigenvalues of H^a with $\#(a) \geq 2$ are called *thresholds*.

* * *

[16] contains extensive discussion of properties of eigenfunctions of $-\Delta + V$. Here we state some of the most important results. For many purposes, the natural class of potentials, V, for this study is K^ν defined by:

Definition Let $\nu \geq 3$; $V \in K^\nu$ if and only if

$$\lim_{\alpha\downarrow 0} \sup_x \int_{|x-y|\leq\alpha} |x-y|^{-(\nu-2)} |V(y)| d^\nu y = 0$$

If $\nu = 2$, K^ν is defined with $|x-y|^{-(\nu-2)}$ replaced by $\ln(|x-y|^{-1})$ and if $\nu = 1$, $V \in K^\nu$ if and only if $\sup_x \int_{|x-y|\leq 1} |V(y)| d^\nu y < \infty$.

$V \in K^\nu$ does *not* imply that $-\Delta + V$ is essentially self-adjoint on C_0^∞, but one can [16] always define a self-adjoint operator "$-\Delta + V$" by a method of forms: This meaning agrees with that obtained by closing the operator on C_0^∞ in case the operator sum is self-adjoint there. The following three results are discussed (either proofs or references given) in [16]. If $Hu = Eu$, then $(-\Delta+(V-E))u = 0$, so by changing V, we can look at functions with $Hu = 0$ and obtain information on general eigenfunctions.

Theorem 1.5 (Sobolev estimates for H) Let $H = -\Delta + V$ with $V_- \equiv \max(-V,0)$ in K^ν and $V_+ = \max(V,0)$ in K^ν_{loc}. Let $k > \nu/4$. Then any function in $D(|H|^k)$ is a bounded continuous function.

Theorem 1.6 (Subsolution estimate for H) Let H obey the hypotheses of Thm. 1.5. Let u obey $Hu = 0$ in distributional sense (u *not* necessarily in L^2). Then

$$|u(x)| \leq C \int_{|x-y|\leq 1} |u(y)| d^\nu y$$

for a constant C depending only on K^ν norms of V_-.

Theorem 1.7 (Harnack's inequality for H) Let $V \in K^\nu_{loc}$. Let Ω be a bounded open set and K compact in Ω. Then, there is a constant C depending only on K^ν norms of $V\chi_\Omega$ so that every solution, u, of $Hu = 0$ in Ω with u non-negative on Ω, obeys

$$C^{-1}u(x) \leq u(y) \leq Cu(x)$$

for all x,y.

We will not indicate in detail the proofs of the last two theorems. In many ways, the key is the study of the Poisson kernel for H, i.e. for a small open ball, B, about a point x, one can study the map, M^B_V from continuous functions f on ∂B to functions on B defined by $M^B_V(f) = u$ obeys $Hu = 0$ in distributional sense on B and $u(x) \to f(y)$ as $x \to y$ on ∂B. It happens that $(M^B_V f)(x) = \int_{\partial B} P^B_V(x,y) f(y) d\omega(y)$. The last two theorems are proven by showing that P is bounded above and away from zero as x runs through a compact subset of B and y runs through ∂B. This is precisely what Aizenman-Simon [17] do. Recently, Zhao [18] and Brossard [19] have actually proven more subtle estimates showing that $P^B_V(x,y)/P^B_{V=0}(x,y)$ is bounded above and below uniformly in x and y (i.e. they show the boundary behavior of P is essentially V independent).

2. Bound state problems

"Bound states" is the name given to eigenfunctions of eigenvalues in the discrete spectrum (isolated points of the spectrum of finite multiplicity). There are various aspects of the study of eigenfunctions: (i) Identify $\sigma_{ess}(H)$ $(=\sigma(H)\backslash\sigma_{disc}(H))$ (ii) Let N denote the sum of the dimensions of the eigenspaces associated to all points in σ_{disc}. Is N finite or infinite? (This is the same as asking if $\#(\sigma_{disc})$ is finite or infinite.) (iii) If N is finite, can one obtain effective bounds on it? (iv) When is $N = 0$?

For two body systems, $-\Delta + V$ with V decaying at ∞, there is a large literature on these questions, summarized in [20]. We will single out two results for special mention, but first we need to find $\sigma_{ess}(-\Delta+V)$ in this case.

Definition Let A be a self-adjoint operator. B is called *A-compact* if and only if $D(B) \supset D(A)$ and $B(A+i)^{-1}$ is compact.

The methods of the proof of Thm. 1.2 imply easily that

Proposition 2.1 If $\nu \leq 3$ and $\lim_{|x|\to\infty} \int_{|y-x|\leq 1} |V(y)|^2 dy = 0$ or if $\nu \geq 4$ and $\lim_{|x|\to\infty} \int_{|x-y|\leq 1} |y-x|^{-(\nu-4+\alpha)} |V(y)|^2 dy = 0$ for some $\alpha > 0$, then V is $-\Delta$-compact.

We write S^{ν}_{comp} for the V's given in Prop. 2.1.

Proposition 2.2 If A is self-adjoint, and if B is A-compact and symmetric, then $\sigma_{ess}(A+B) = \sigma_{ess}(A)$.

Proof A simple theorem of Weyl (see [3]) says that $E \in \sigma_{ess}(C)$ if and only if there exists a sequence of vectors $\varphi_n \in D(C)$ with $\varphi_n \to 0$ *weakly* and $\|(C-E)\varphi_n\| \to 0$, $\|\varphi_n\| \to 0$. Given $E \in \sigma_{ess}(A)$, find such a sequence, let $\psi_n = (E^2+1)(A^2+1)^{-1}\varphi_n$. It is not hard to show that $\psi_n \to 0$ weakly, $\|(A+B-E)\psi_n\| \to 0$, $\|\psi_n\| \to 1$. Thus, $E \in \sigma_{ess}(A+B)$ and we conclude that $\sigma_{ess}(A) \subset \sigma_{ess}(A+B)$. Using $(A+B+i)^{-1} = (A+i)^{-1}(1+B(A+i)^{-1})^{-1}$, one can show that B is $(A+B)$-compact. Thus, $\sigma_{ess}(A+B) \subset \sigma_{ess}(A)$ by repeating the above argument. ∎

Corollary 2.3 If V lies in S^{ν}_{comp}, then $\sigma_{ess}(-\Delta+V) = [0,\infty)$.

We return now to N for $-\Delta + V$ which we denote by $N(V)$. We want to single out two results:

Theorem 2.4 (Quasiclassical bounds on N(V)) Let $\nu \geq 3$. There is a universal constant C_ν so for all $V \in L^{\nu/2}$,

$$N(V) \leq C_\nu \int |V(x)|^{\nu/2} d^\nu x$$

This theorem is particularly important because the semiclassical approximation for $N(V)$ is to take the volume in phase space where $p^2 + V(x)$ is negative and divide it by $(2\pi)^\nu$ (for $\hbar=1$, so h is 2π). Thus if $V(x) \leq 0$:

$$N_{c\ell}(V) = \frac{\tau_\nu}{(2\pi)^\nu} \int |V(x)|^{\nu/2} d^\nu x$$

where τ_ν is the volume of the unit sphere in R^ν. As a result, Thm. 2.3 says that the quantum $N(V)$ is bounded by a multiple of $N_{c\ell}(V)$. There is also a connection with Sobolev estimates (see [21,22]). Thm. 2.1 was proven independently (with different C_ν) by Rosenbljum [23], Cwickel [24] and Lieb [25] (see [21,26] for expositions of [25,24]) with newer proofs by Li-Yau [27] and Fefferman-Phong [28]. Theorem 2.4 is in some sense especially accurate for "large" V:

Theorem 2.5 (Quasiclassical limit for N(V)). Let $\nu \geq 3$, $V \in L^{\nu/2}$. Then

$$\lim_{\lambda \to \infty} N(\lambda V)/N_{c\ell}(\lambda V) = 1$$

Since $-\Delta + \lambda V = (-\lambda^{-1}\Delta + V)\lambda$, the $\lambda \to \infty$ limit is "equivalent" to the $\hbar \to 0$ limit, which is "why" the semiclassical result is asymptotically correct. Thm. 2.4 is used to show that Thm. 2.5 need only be proven when $V \in C_0^\infty$ where Thm. 2.5 was proven independently by Birman-Borzov [29], Martin [30] and Tamura [31] (see [3,21] for pedagogic discussions). A multiparticle analog of Thm. 2.4 can be found in [32].

* * *

Now we want to describe some results on bound states for multiparticle systems. The first basic result describes $\sigma_{ess}(H)$. We first use the partition notation described in Section 1.

Theorem 2.6 (HVZ theorem) Let H be the Hamiltonian (with C.M. motion removed) of an N-body system on $L^2(R^{\mu(N-1)})$ with two body potential in S^{μ}_{comp}. Let

$$\Sigma = \inf_{a|\#(a)\geq 2} [\min \sigma(H(a))]$$

Then $\sigma_{ess}(H) = [\Sigma, \infty)$.

In order to understand this result, it is useful to know

Theorem 2.7 (Persson's theorem [33]) Let $V_- \in K^{\nu}$, $V_+ \in K^{\nu}_{loc}$. Then

$$\inf \sigma_{ess}(-\Delta+V) = \lim_{R\to\infty} \inf\{(\varphi,(-\Delta+V)\varphi) | \varphi\in C_0^{\infty}(R^{\nu}); \|\varphi\|=1; \operatorname{supp}\varphi\subset\{x||x|>R\}\}$$

For a proof, see also Agmon [34,35] or Cycon et al. [4]. What Persson's theorem suggests is that essential spectrum is associated with vectors living near infinity (this is basically because $(H+i)^{-1}$ times the characteristic function of a bounded set is compact). Thus, in the N-body case, essential spectrum is associated with states near infinity where the system must break up into two or more subsets. Thus, one should expect

$$\sigma_{ess}(H) = \bigcup_{a|\#(a)\geq 2} \sigma(H(a))$$

which is just a restatement of Thm. 2.6.

Thm. 2.6 has two parts in a natural sense: (i) $[\Sigma,\infty) \subset \sigma(H)$ and (ii) $\sigma(H) \cap (-\infty,\Sigma)$ is discrete. (i) is the "easy" half and (ii) will be what we concentrate on (see e.g. Garding [36] for the "easy" half). The name HVZ recognizes contributions of Hunziker [37], van Winter [38] and Zhislin [39]. Zhislin used geometric ideas together with rather extensive machinery, so for some years the integral equation proof of van Winter and Zhislin was considered the more elementary (see e.g. [3] for that proof), but with the work of Enss [40] and Simon [41], the geometric proof has come into fashion, and it is Sigal's version of it [42] that we will sketch.

We begin with a basic result on localization called the "IMS localization formula" due to contributions of Ismigilov, Morgan, Simon and I.M. Sigal, who first appreciated its great usefulness.

Proposition 2.8 Let $\{j_a\}$ be a finite family of functions with distributional gradients in L^∞ obeying $\sum j_a^2 = 1$. Let $H = -\Delta + V$ on $L^2(R^\nu)$ have C_0^∞ as a form core. Then

$$H = \sum j_a H j_a - \sum (\nabla j_a)^2 \qquad (2.1)$$

Remark (2.1) is intended in the sense of expectation values with $(\varphi, j_a H j_a \varphi) \equiv (j_a\varphi, H j_a \varphi)$. If the j's are sufficiently smooth, it holds in operator sense.

Proof By a limiting argument, we can suppose the j's are C^∞. Then $[j_a,[j_a,H]] = [j_a,[j_a,-\Delta]] = -2(\nabla j_a)^2$. Thus

$$\sum_a j_a^2 H + H j_a^2 = 2\sum_a j_a H j_a - 2\sum_a \nabla j_a^2$$

which yields (2.1) given $\sum j_a^2 = 1$. ■

Next, we need the existence of a special partition of unity for N-body system: A *Ruelle-Simon partition of unity* of an N-body system is a set of functions $\{j_a\}$ on X (the CM=0 space) labeled by partitions, a, with $\#(a) = 2$ obeying (i) j_a is C^∞ (ii) $\sum j_a^2 = 1$ (iii) if $\lambda > 1$ and $\|x\| > 1$, then $j_a(\lambda x) = j_a(x)$ (iv) for some $C > 0$,

$$[\operatorname{supp} j_a \cap \{x \mid \|x\| > 1\}] \subset \bigcup_{(ij)\not\subset a} \{x \mid \|x_i - x_j\| \geq C\|x\|\}$$

Thus j_a lives in the region where particles in different clusters of a are far from each other as $\|x\| \to \infty$. (The norm of x is measured in any convenient way.)

Proposition 2.9 Ruelle-Simon partitions of unity exist.

Sketch of proof Let $S_a = \{x \mid \|x\| = 1,\ x_i = x_j \text{ for some } (ij) \subset a\}$. We claim that $\bigcap_a S_a = \phi$. For if $\|x\| = 1$, $x_i \neq x_j$ for some pair i,j (since $\sum m_i x_i = 0$). Let $a = \{C_1, C_2\}$ with $C_1 = \{x \mid x = x_i\}$ and $C_2 = \{1,\dots,N\} \setminus C_1$. Then $x \notin S_a$. Since the S_a are closed and $\bigcap_a S_a = \phi$, it is not hard to find $C^\infty \tilde{j}_a$ on $\{x \mid \|x\| = 1\}$ so that $\sum \tilde{j}_a^2 = 1$ and $\tilde{j}_a$ vanishes in a neighborhood of S_a. Now let $j_a(x) = \tilde{j}_a(x/\|x\|)$ if $\|x\| \geq 1$ and continued to be smooth near 0. ■

One can actually estimate the constant C; see [41].

Proposition 2.10 Let H be an N-body Hamiltonian of the type described in Thm. 2.6, and let $\{j_a\}$ be a Ruelle-Simon partition of unity. Then:

(a) $(\nabla j_a)^2$ is H-compact for any a

(b) $I(a)j_a$ is H-compact for any a.

Proof $(\nabla j_a)^2$ is bounded and goes to zero at ∞ (as $|x|^{-2}$) so (a) is easy. If $(ij) \not\subset a$, $|x_i - x_j| \to \infty$ as $|x| \to \infty$, so it is not hard to show that $V_{ij}j_a$ is H-compact (see e.g. [41]). ■

We are now ready for

Proof of Thm. 2.6 [42] (Hard direction) Pick a Ruelle-Simon partition of unity. Write

$$H = \hat{H} + W, \quad \hat{H} = \sum_{\#(a)=2} j_a H(a) j_a$$

$$W = \sum j_a I(a) j_a - \sum (\nabla j_a)^2$$

W is H-compact, so $\sigma_{ess}(H) = \sigma_{ess}(\hat{H})$ by Prop. 2.2. Since $H(a) \geq \Sigma$ for all a, we have for any φ in L^2 that

$$(\varphi, \hat{H}\varphi) \geq \sum_a (j_a^2\varphi, \Sigma\varphi) = \Sigma (\varphi, \varphi)$$

so $\sigma_{ess}(\hat{H}) \subset \sigma(H) \subset [\Sigma, \infty)$. ■

* * *

These geometric methods have been extremely useful in studying bound state problems in N-body systems. Here are some results which we quote without detailed proofs:

Theorem 2.11 Let $\nu \geq 3$. Let H be the Hamiltonian of an N-body system with potentials V_{ij} obeying $|V_{ij}(x)| \leq C(1+|x|)^{-2-\varepsilon}$. Suppose that the bottom of the continuum is two body in the sense that

$$\inf_{\#(a)\geq 3} \sigma(H(a)) > \inf_{\#(a)\geq 2} \sigma(H(a))$$

Then H has only a finite discrete spectrum.

Theorems of this genre go back to Zhislin and collaborators, see e.g. [43,44]. This result is proven by geometric means in Sigal [42].

Theorem 2.12 For any N,Z let $H(N,Z)$ be the operator on $L_a^2(R^{3N})$ (≡ function on $R^{3N} = \{x=(x_1,\dots,x_N)|x_i \in R^3\}$ antisymmetric in the x_i's) given by

$$H(N,Z) = \sum_{i=1}^{N} -\Delta_i - \frac{Z}{|x_i|} + \sum_{i\neq j} \frac{1}{|x_i - x_j|}$$

Let $E(N,Z) \equiv \inf \operatorname{spec} H(N,Z)$. Then, there exists $N(Z)$, so that

$$E(N+1,Z) = E(N,Z) \qquad \text{if } N \geq N(Z)$$

This result says that a nucleus of charge Z bonds at most $N(Z)$ electrons (we will take $N(Z)$ to be the smallest $N(Z)$ obeying the above). Thm. 2.12 with L_a^2 replaced by L^2 was proven by Ruskai [45]; Thm. 2.12 was proven by Sigal [42]. Recently, Lieb [46] has found an elegant direct proof that $N(Z) \leq 2Z$ for all Z. Using Sigal's method, Lieb et al. [47] have proven that $\lim_{Z\to\infty} N(Z)/Z = 1$.

3. The basic notions of scattering theory

We will introduce here some of the simplest notions in scattering theory; Enss, in his lectures, will discuss much more involved ideas. See [2] for an extensive discussion of scattering theory.

Given A,B, we want to find pairs φ,ψ so

$$e^{-itA}\varphi - e^{-itB}\psi \to 0 \quad \text{as} \quad t \to +\infty \tag{3.1}$$

It turns out that for general B, one shouldn't normally expect that a φ exists for every ψ. For example, if $B\psi = 0$, one must have $\varphi = \psi$ and $A\psi = 0$ for the limit (3.1) to occur. Thus, we only try to find φ for $\psi \in \mathcal{H}_{a.c.}(B)$, the absolutely continuous subspace for B. Note (3.1) is equivalent to

$$\varphi = \lim_{t\to\infty} e^{itA}e^{-itB}\psi \tag{3.2}$$

This motivates

Definition Given two self-adjoint operators, A,B we say that the *wave operators* $\Omega^{\pm}(A,B)$ exist if and only if $\operatorname{s-lim}_{t\to\mp\infty} e^{itA}e^{-itB}P_{a.c.}(B)$ exists.

Notice that if s is fixed and we replace t by $t-s$, the limit is the same. Thus:

$$e^{-isA}\Omega^{\pm}(A,B) = \Omega^{\pm}(A,B)e^{-isB} \tag{3.3}$$

This implies that B restricted to $\operatorname{Ran} P_{a.c.}(B)$ and A restricted to $\operatorname{Ran} \Omega^{\pm}(A,B)$ are unitarily equivalent. In particular, $\operatorname{Ran} \Omega^{\pm}(A,B) \subset \operatorname{Ran} P_{a.c.}(A)$. It is clearly natural to single out:

Definition Let $\Omega^{\pm}(A,B)$ exist. We say they are *complete* if and only if $\operatorname{Ran} \Omega^{\pm}(A,B) = \operatorname{Ran} P_{a.c.}(A)$.

If $\Omega^{\pm}$ exist and are complete, then the association (3.1) sets up a one-one correspondence between $\mathcal{H}_{a.c.}(A)$ and $\mathcal{H}_{a.c.}(B)$. Given the fact that (3.1) is symmetric in A and B, it is not hard to show:

Proposition 4.1 Let $\Omega^{\pm}(A,B)$ exist. Then, they are complete if and only if

$\Omega^{\pm}(B,A)$ exist.

Remark Deift-Simon [48] have an N-body analog of Prop. 4.1.

Proposition 4.1 suggests that one look for a condition symmetric in A,B which implies that $\Omega^{\pm}(A,B)$ exist. The strongest such result seems to be:

Theorem 4.2 Let A,B be self-adjoint operators with $(A+i)^{-1}-(B+i)^{-1}$ compact and so that for any interval, Δ: $E_{\Delta}(A)(A-B)E_{\Delta}(B)$ is trace class (where $E_{\Delta}(\cdot)$ is a spectral projection). Then $\Omega^{\pm}(A,B)$ exist and are trace class.

Theorems of this genre go back to Kato and Rosenbljum, with later contributions of note by Kato, Birman, Pearson and Davies. This result follows from an observation of Davies [49] and a theorem of Birman which appears as Cor. 6.7 in [26].

As far as the abstract theory is concerned, Thm. 4.2 is quite elegant. However, in the concrete situation of $A=-\Delta+V$ and $\Omega=-\Delta$ on $L^2(R^{\nu})$, it breaks down when V decays more slowly than $|x|^{-\nu}$ while one expects that $\Omega^{\pm}(A,B)$ and are complete so long as V only has $|x|^{-1-\varepsilon}$ decay. Various methods exist for proving that this compactness result (existence is easy, see [2], Sect. XI.4 and references therein):

(a) The method of weighted L^2 estimates developed by Agmon and Kuroda and discussed in Section XIII.8 of [3].

(b) The Enss method discussed in Section XI.17 of [2], and the book of Perry [50].

(c) Combining the Mourre estimates, to be described in the next section, with the smoothness techniques of Kato and Lavine (see e.g. Perry, Sigal, Simon [51]).

For N-body systems, the notion of completeness requires a more elaborate definition. Let H^a be the Hamiltonian describing internal motion for the clustering a and let P^a denote the projection onto the span of the eigenvectors of H^a, and let $P(a)=P^a\otimes I$. One defines for any a

$$\Omega_a^{\pm}=\lim_{t\to\mp\infty} e^{itH}e^{-itH(a)}P(a)$$

Under suitable hypotheses, it is not hard to show that $\Omega_a^{\pm}$ exist (see Thm. XI.35 in [2], but note the arguments there require modification if $\nu=1,2$). $\Omega_a^{\pm}\varphi=\eta$ is a state with $e^{-itH}\eta$ asymptotic as $t \to -\infty$ to a state with *bound* clusters in a moving relatively freely. It is thus reasonable and not hard to prove ([2], Thm. XI.36(b)) that

$$\operatorname{Ran} \Omega_a^{+} \perp \operatorname{Ran} \Omega_b^{+} \qquad a \neq b$$

Completeness now reads

$$L^2(R^{(N-1)\nu}) = \underset{a}{\oplus}(\operatorname{Ran} \Omega_a^{\pm}) \tag{3.4}$$

Notice that if a_1 is the unique clustering with $\#(a_1) = 1$ (so $H(a_1) = H$), then $\Omega_{a_1}^{\pm} = P(a_1)$ is the projection onto the point spectral subspace H, and that by the intertwining relation

$$e^{itH}\Omega_a^{\pm} = \Omega_a^{\pm} e^{itH(a)}$$

$\operatorname{Ran} \Omega_a^{\pm} \subset \mathcal{H}_{a.c.}(H)$ if $\#(a) \geq 2$. Thus (3.4) implies that H has empty singular continuous spectrum.

Thus far, there are fairly general results on three body completeness [52,53,54] but only very specialized results for N-body, see e.g. [55,56,57,58]. It has been emphasized to me by I.M. Sigal that the following result which extends an idea of Deift-Simon [48] should be very useful in a possible inductive proof of completeness:

<u>Proposition 4.3</u> Let $\{A(a)\}_{\#(a)\geq 2}$ be bounded operators with $\sum_a A(a) = 1$. Suppose that (i) Each H^a with $\#(a)\geq 2$ is complete (ii) The operators $\lim_{t\to\mp\infty} e^{itH(a)}A(a)e^{-itH}P_{a.c.}(H) \equiv W_a^{\pm}$ exist. (iii) $\mathcal{H}_{sing}(H) = \phi$. Then H is complete.

<u>Proof</u> Let $\eta \in \operatorname{Ran} P_{a.c.}(H)$. Then

$$e^{-itH}\eta = \sum_a A(a)e^{-itH}\eta \sim \sum_a e^{-itH(a)}W_a^{\pm}\eta$$

where $\sim$ means the difference goes to zero as $t \to \mp\infty$. Since H^a is complete, $e^{-itH(a)}\varphi$ is asymptotically a sum of vectors of the form $e^{-itH(b)}P(b)\varphi_b$ with $b \leq a$, so we have completeness for H. ∎

4. Mourre estimates

Eric Mourre, in a deep paper [59], singled out certain estimates which he showed have important spectral consequences, and which he proved for a large class of two and three body systems. Perry, Sigal, Simon [51] then gave an involved proof of these estimates for general N-body systems. Subsequently, Froese-Herbst [80] found a rather simple proof of these PSS results.

Let H be a self-adjoint operator, and A a second self-adjoint operator. We will not be explicit about all domains referring the reader to [59,51] for explicit hypotheses. Under such hypotheses, one can define $-i[A,H] = B$ originally on a suitable core for H and then as an "operator" from $D(H)$ to $D^{-1}(H)$ (equal abstract dual of $D(H)$) i.e. $(H+i)^{-1}B(H+i)^{-1}$ is a bounded operator. We say that H obeys a *Mourre estimate* at a point E_0, if there is an open interval Δ about E_0 so that

$$E_\Delta B E_\Delta \geq \alpha E_\Delta^2 + E_\Delta K E_\Delta \tag{4.1}$$

for a *compact* operator K, and some $\alpha > 0$.

Theorem 4.1 Under suitable domain hypotheses (including a bound on $[A,B]$), if a Mourre estimate holds for any $E_0 \in I$, an open interval, then

(i) H has no singular continuous spectrum in I

(ii) In any compact $J \subset I$, there are finitely many eigenvalues of H and each eigenvalue has finite multiplicity.

(iii) For any compact $J \subset I$ and $\delta > 0$, $\sup_{\substack{0<\varepsilon<1 \\ E\in J}} \|(|A|+1)^{-\frac{1}{2}-\delta}(H-E-i\varepsilon)^{-1}(|A|+1)^{-\frac{1}{2}-\delta}\|$ $< \infty$.

The result is essentially due to Mourre [59], although the above include refinements of [51]. While we will not give the proof in detail, we note the basic idea behind (ii). Using the unstated domain conditions, one verifies a Virial theorem: If $H\varphi = E\varphi$, then $(\varphi, B\varphi) = 0$. Thus, if $H\varphi_n = E_n\varphi_n$ with $E_n \to E_0 \in \Delta$ and φ_n is orthonormal, then, by (4.1)

$$0 \geq \alpha\|\varphi_n\|^2 + (\varphi_n, K\varphi_n)$$

which is impossible since K is compact and $\varphi_n \to 0$ weakly.

Mourre [61] (see also [62]) has also proven interesting propagation estimates when Mourre estimates hold.

When does an estimate like (4.1) hold? Mourre had the idea of taking $A = \frac{1}{2}(x \cdot p + p \cdot x)$ (with $p = -i\nabla$), the generator of dilations. For two body operators, $H = H_0 + V$

$$-i[A,H] = 2H_0 - x \cdot \nabla V = 2H + W$$

where

$$W = -x \cdot \nabla V - 2V$$

If $K = E_\Delta(H) W E_\Delta(H)$ is compact, and if $\Delta = [a,b]$ with $a > 0$, then $E_\Delta B E_\Delta \geq 2aE_\Delta^2 + E_\Delta K E_\Delta$, so a Mourre estimate holds.

<u>Proposition 4.2</u> If $V = V_1 + V_2$ with $V_1(H_0+i)^{-1}$, $x \cdot \nabla V_1(H_0+i)^{-1}$ and $(1+|x|)V_2(H_0+i)^{-1}$ all compact, then a Mourre estimate holds for $A = \frac{1}{2}(x \cdot p + p \cdot x)$; $H = -\Delta + V$ and $E_0 > 0$.

<u>Proof</u> Note first that $D(H) = D(H_0)$, so $(H_0+i)E_\Delta(H)$ is bounded. Thus, $E_\Delta W_1 E_\Delta$ is obviously compact as is $E_\Delta V_2 E_\Delta$. As for $-i[A,V_2]$, we can write that as $-\sum_i [\nabla_i, x_i V_2] + \nu V_2$ and $E_\Delta[(\nabla_i)(x_i V_2)]E_\Delta$ is compact. ■

Mourre [59] showed how to do this for three-body systems and then PSS [51] proved:

<u>Theorem 4.3</u> If each $V_{ij} = V_{ij}^{(1)} + V_{ij}^{(2)}$, where as operators on $L^2(R^\nu)$, $V^{(1)}(h_0+i)^{-1}$, $x \cdot \nabla V^{(1)}(h_0+i)^{-1}$ and $(1+|x|)V^{(2)}(h_0+i)^{-1}$ are all compact on ($h_0 = -\Delta$ on R^ν), then a Mourre result holds for any $E_0 \neq$ threshold (and α is twice the distance from E_0 to the threshold of next lowest energy).

To conclude (i) and (iii) in Thm. 4.1, we also need control on $[A,B]$ which requires V_{ij} have more decay than in the above theorem (e.g. $(1+x^2)V^{(2)}(h_0+i)^{-1}$ $V^{(1)}(h_0+i)^{-1}$ and $x^2\nabla\nabla V^{(1)}(h_0+i)^{-1}$ compact will do); see e.g. [51]. Froese-Herbst [60] have a simple proof of Thm. 4.3.

Froese-Herbst [63] have proven the following theorems using Mourre estimates:

Theorem 4.4 [63] Let V_{ij} obey the hypotheses of Theorem 4.3, and suppose that $H\varphi = E\varphi$, with $\varphi \in L^2$. Let $|x|$ denote the norm of x in X (i.e. $(\sum m_i x_i^2)^{\frac{1}{2}}$ if $\sum m_i x_i = 0$). Define

$$\alpha = \sup\{a \mid e^{a|x|}\varphi \in L^2\}$$

Then either $\alpha = \infty$ or $\alpha^2 + E$ is a threshold.

By using results [64] which imply $\alpha = \infty$ is not allowed:

Theorem 4.5 [63] If the V_{ij} obey the hypotheses of Theorem 4.3, and for all $\varepsilon > 0$

$$y \cdot \nabla V_{ij}(y) \leq \varepsilon h_0 + C_\varepsilon$$

then $H\varphi = E\varphi$ has no L^2 solutions with $E > 0$.

5. An Introduction to the Theory of Stochastic Jacobi Matrices

In this final section, we consider another topic currently of intense interest, namely Schrödinger operators with random or almost periodic potentials. For technical simplicity, we will restrict ourselves to $\nu = 1$ and we will discretize space, i.e. replace R by Z and $-d^2/dx^2$ by a second difference operator. See [65] for an extensive bibliography including papers dealing with the continuum case and with $\nu > 1$.

We should take h_0 to be the finite difference analog of $-d^2/dx^2$, namely $(h_0u)(n) = \delta^{-2}[2u(n) - u(n+1) - u(n)]$. First of all, we take $\delta = 1$ for convenience. Then, we replace h_0 by $h_0 - 2$ which won't change any spectral properties. Then we make the unitary transformation $u(n) \to (-1)^n u(n)$ which means that instead, we take

$$(h_0u)(n) = u(n+1) + u(n-1) \tag{5.1}$$

on $\ell^2(Z)$. We will study not individual operators but whole classes: Let (Ω,μ) be a probability measure space and let $T : \Omega \to \Omega$ be an invertible, measure preserving ergodic transformation. Let $f : \Omega \to R$ be a bounded measurable function. Given $\omega \in \Omega$, define

$$V_\omega(n) = f(T^\alpha \omega) \tag{5.2}$$

and

$$h_\omega = h_0 + V_\omega \tag{5.3}$$

We ask about properties of h_ω that hold for a.e. ω.

<u>Examples</u> 1. $\Omega = \displaystyle\mathop{X}_{n=-\infty}^{\infty} [a,b]$, $d\mu = \displaystyle\bigotimes_{n=-\infty}^{\infty} d\nu(x_n)$ where $d\nu$ is a probability measure on $[a,b]$. Let $(Tx)_n = x_{n+1}$ and $f(x) = x_0$. Then the variables $V_\omega(n)$ are precisely independent identically distributed (i.i.d.) random variables with common density $d\nu$. This is conventionally called "random potentials." The case $d\nu(x) = (b-a)^{-1}\chi_{(a,b)}(x)dx$ is called *the Anderson model*.

2. Let Ω be the k torus $\{(\theta_1,\ldots,\theta_k); 0 \le \theta < 1\}$ with its structure as a group (addition of components, mod. 1) and Haar measure $\prod_1^k d\theta_i$. Let f be

a continuous function on Ω and let $(T\theta)_i = \theta_i + \alpha_i \pmod 1$ where $\alpha_1, \ldots, \alpha_k$ are numbers so that $1, \alpha_1, \ldots, \alpha_k$ are independent over the rationals. Then $V_\theta(n) = f(\alpha_i n + \theta_i)$ is quasiperiodic. A good example is $V_\theta(n) = \lambda \cos(2\pi\alpha n + \theta)$ (now θ runs through $[0, 2\pi)$) called *Happer's equation* or the *almost Mathieu equation*. An interesting example (see [66,67,68]) which doesn't quite fit into this framework is $V_\theta(n) = \lambda \tan(\pi\alpha n + \theta)$. This is called the *Maryland model* and has the feature of being exactly soluble in a certain sense.

It makes sense to study the totality of the operators $\{h_\omega\}$ for one has the following consequence of ergodicity.

Theorem 5.1 ([69]) The following sets are constant for a.e. ω (i.e. there is a set $S \subset \Omega$ whose complement has measure zero, so that if $\omega, \omega' \in S$ all the objects below are equal for ω and ω'): $\sigma(h_\omega)$, $\sigma_{a.c.}(h_\omega)$, $\sigma_{pp}(h_\omega)$ ($\equiv$ *closure* of set of eigenvalues), $\sigma_{s.c.}(h_\omega)$. Moreover, $\sigma_{disc}(h_\omega) = \phi$ and $\sigma(h_\omega)$ has *no* isolated points.

Remark $\sigma_{disc}(h_\omega) = \phi$ also in the higher dimensional case; it is also true in that case that $\sigma(h_\omega)$ has no isolated points, but this is more subtle (see [70,71]).

Here are some typical results illustrating the subtle spectral properties of stochastic Jacobi matrices:

Theorem 5.2 Let h_ω have a random potential ($V_\omega(n)$ i.i.d.'s) with $d\nu(x) = F(x)dx$ (supported on $[a,b]$). Then, for a.e. ω,

$$\mathrm{spec}(h_\omega) = [-2,2] + \mathrm{supp}(F)$$

and h_ω has a complete set of eigenfunctions.

For proofs see [69,72]. For related continuum results, see [73,74]. For the study of $h_0 + (1+|n|)^{-\alpha} V_\omega(n)$, see [75,76].

Theorem 5.3 Let $\{a_n\} \in \ell_1(0,1,\ldots)$ and let $h(a_m) = h_0 + \sum_{m=0}^{\infty} a_m \cos(2\pi n/2^m)$. Then for a dense G_δ in ℓ_1, $h(a_m)$ has a nowhere dense spectrum and for a dense set in ℓ_1, $\sigma(h(a_m))$ is both nowhere dense and purely absolutely continuous.

See [77,78,79] for proofs; see [80] for a discussion of nowhere dense a.c. spectrum.

Theorem 5.4 Pick any $0 < \alpha < 1$. Then, there exists almost periodic potentials $V_\omega(n)$ so that $h_0 + V_\omega(n) = h_\omega$ has dense point spectrum and $\sigma(h_\omega)$ has Hausforff dimension α.

The basic idea is from Craig [81], although his examples are not strictly almost periodic; those are due to Poschel [82]. See also [83].

Sarnak [84] first suggested that spectral properties should depend on Diophantine properties of α:

Theorem 5.5 Let α be an irrational number for which there exist rational approximations p_n/q_n obeying $|\alpha - p_n/q_n| \leq n^{-q_n}$. Let $\lambda > 2$. Then

$$h_0 + \lambda \cos(2\pi\alpha n+\theta)$$

has purely singular continuous spectrum.

For a proof, see Avron-Simon [85]; important input comes from Aubry-André [86] and Gordon [87]. The set of α obeying the estimates is a dense G_δ in R (of Lebesgue measure zero).

Definition A stochastic process $V_\omega(n)$ is called *deterministic* if and only if $\{V_\omega(n)\}_{n\geq 0}$ is (a.e.) a measurable function of $\{V_\omega(n)\}_{n<0}$. For example, a.p. functions yield deterministic processes; random potentials do not.

Theorem 5.6 If h_ω is a stochastic Jacobi matrix and h_ω has some a.c. spectrum (for a.e. ω), then V_ω is a deterministic process.

This result in the continuum case is due to Kotani [88]; see Simon [89] for the discrete case.

References

1. M. Reed and B. Simon, *Methods of modern mathematical physics, Vol. II. Fourier analysis, self adjointness,* Academic Press, 1975.

2. M. Reed and B. Simon, *Methods of modern mathematical physics, Vol. III. Scattering theory,* Academic Press, 1978.

3. M. Reed and B. Simon, *Methods of modern mathematical physics, Vol. IV. Analysis of operators,* Academic Press, 1977.

4. H. Cycon, R. Froese, W. Kirsch and B. Simon, *Lectures on Schrödinger operators,* in preparation.

5. M. Reed and B. Simon, *Methods of modern mathematical physics, Vol. I. Functional analysis,* Academic Press, 1972.

6. T. Kato, *Perturbation theory for linear operators,* Springer, 1966.

7. M. Schechter, *Spectra of partial differential operators,* North Holland, 1971.

8. T. Kato, Schrödinger operators with singular potentials, Is. J. Math. 13 (1973), 135-148.

9. B. Simon, Maximal and minimal Schrödinger forms, J. Op. Th. 1 (1979), 37-47.

10. H. Leinfelder and C. Simader, Schrödinger operators with singular magnetic vector potentials, Math. Z. 176 (1981), 1-19.

11. F. Rellich, Störungstheorie der spectralzerlegung, II, Math. Ann. 116 (1939), 555-570.

12. T. Kato, Fundamental properties of Hamiltonian operators of Schrödinger type, Trans. Am. Math. Soc. 70 (1951), 195-211.

13. F. Stummel, Singulare elliptische differentialoperatoren in Hilbertschen Räumen, Math. Ann. 132 (1956), 150-176.

14. A.G. Sigalov and I.M. Sigal, Description of the spectrum of the energy operator of quantum-mechanical systems..., Theor. and Math. Phys. 5 (1970), 990.

15. P. Deift, W. Hunziker, B. Simon and E. Vock, Pointwise bounds on eigenfunctions and wave packets in N-body quantum systems, IV, Commun. Math. Phys. 64 (1978), 1-34.

16. B. Simon, Schrödinger semigroups, Bull. AMS 7 (1982), 447-526.

17. M. Aizenman and B. Simon, Brownian motion and Harnack's inequality for Schrödinger operators, Comm. Pure Appl. Math. 35 (1982), 209-273.

18. Z. Zhao, Uniform boundedness of conditional gauge and Schrödinger equations, Commun. Math. Phys., to appear, and Continuity of conditioned Feynman-Kac functional and integral kernel for Schrödinger equation, Stanford preprint.

19. J. Brossard, The problem of Dirichlet for the Schrödinger operator, Institut Fourier preprint, 1984.

20. B. Simon, On the number of bound states of two-body Schrödinger operators - A review, from *Studies in mathematical physics, essays in honor of Valentine Bargmann*, Princeton Press, 1976, pp. 305-326.

21. B. Simon, *Functional integration and quantum physics*, Academic Press, 1979.

22. E.H. Lieb and W. Thirring, Inequalities for the moments of the eigenvalues of the Schrödinger Hamiltonian and their relation to Sobolev inequalities, in *Studies in mathematical physics, essays in honor of Valentine Bargmann*, Princeton Press, 1976, pp. 269-304.

23. G. Rosenbljum, The distribution of the discrete spectrum for singular differential operators, Dokl. Akad. Nauk SSSR 202 (1972); transl. Soviet Math. Dokl. 13 (1972), 245-249.

24. M. Cwikel, Weak type estimates and the number of bound states of Schrödinger operators, Ann. Math. 106 (1977), 93-102.

25. E. Lieb, Bounds on the eigenvalues of the Laplace and Schrödinger operators, Bull. AMS 82 (1976), 751-753, and Proc. 1979 AMS Honolulu Conference.

26. B. Simon, *Trace ideals and their applications*, Cambridge Univ. Press, 1979.

27. P. Li and S.T. Yau, On the Schrödinger equation and the eigenvalue problem, Commun. Math. Phys. 88 (1983), 309.

28. C. Fefferman, The uncertainty principle, Bull. Am. Math. Soc. 9 (1983), 129.

29. M. Birman and V.V. Borzov, On the asymptotics of the discrete spectrum of some singular differential operators, Topics in Math. Phys. 5 (1972), 19-30.

30. A. Martin, Bound states in the strong coupling limit, Helv. Phys. Acta. 45 (1972), 140-148.

31. H. Tamura, The asymptotic eigenvalue distribution for nonsmooth elliptic operators, Proc. Japan Acad. 50 (1974), 19-22.

32. M. Klaus and B. Simon, Coupling constant threshold in non-relativistic quantum mechanics, II. Two body thresholds in N-body systems, Commun. Math. Phys. 78 (1980), 153-168.

33. A. Persson, Bounds for the discrete part of the spectrum of a semibounded Schrödinger operator, Math. Semd. 8 (1960), 143-153.

34. S. Agmon, *Lectures on exponential decay of solutions of second order elliptic equations. Bounds on eigenfunctions of N-body Schrödinger operators*, Mathematical Notes, Princeton Univ. Press, Princeton, NJ 1982.

35. S. Agmon, On exponential decay of solutions of second order elliptic equations in unbounded domains, Proc. A. Pleijel Conf.

36. L. Garding, On the essential spectrum of Schrödinger operators, J. Func. Anal. 52 (1983), 1-10.

37. W. Hunziker, On the spectra of Schrödinger multiparticle Hamiltonians, Helv. Phys. Acta 39 (1966), 451-462.

38. C. van Winter, Theory of finite systems of particles, I. Mat. Fys. Skr. Danske Vid. Selsk 1 (1964), 1-60.

39. G. Zhislin, Discussion of the spectrum of the Schrödinger operator for systems of many particles, Trudy. Mosk. Mat. Obs. 9 (1960), 81-128.

40. V. Enss, A note on Hunziker's theorem, Commun. Math. Phys. 52 (1977), 233.

41. B. Simon, Geometric methods in multiparticle quantum systems, Commun. Math. Phys. 55 (1977), 259-274.

42. I.M. Sigal, Geometric methods in the quantum many-body problem. Nonexistence of very negative ions, Commun. Math. Phys. 85 (1982), 309-324.

43. G.M. Zhislin, On the finiteness of the discrete spectrum of the energy operator of negative atomic and molecular ions, Theor. Math. Phys. 7 (1971), 571-578.

44. M.A. Antonets, G.M. Zhislin and J.A. Sheresherskii, On the discrete spectrum of the Hamiltonian of an n-particle quantum system, Theor. Math. Phys. 16 (1973), 800.

45. M.B. Ruskai, Absence of discrete spectrum in highly negative ions, I,II. Commun. Math. Phys. 82 (1982), 457; 85 (1982), 325.

46. E.H. Lieb, Bound on the maximum negative ionization of atoms and molecules, Phys. Rev. A, to appear.

47. E.H. Lieb, I.M. Sigal, B. Simon and W. Thirring, Asymptotic bulk neutrality of large Z ions, Phys. Rev. Lett., to appear.

48. P. Deift and B. Simon, A time-dependent approach to the completeness of multiparticle quantum systems, Commun. Pure Appl. Math. 30 (1977), 573-583.

49. E.B. Davies, On Enss' approach to scattering theory, Duke Math. J. 47 (1980), 171-185.

50. P. Perry, *Scattering theory by the Enss method*, Mathematical Reports, Vol. I, 1983.

51. P. Perry, I.M. Sigal and B. Simon, Spectral analysis of multiparticle Schrödinger operators, Ann. Math. 114 (1981), 519-567.

52. L. Faddeev, *Mathematical aspects of the three body problem in quantum scattering theory*, Steklov Institute, 1963.

53. V. Enss, Topics in scattering theory for multiparticle quantum mechanics, to

appear in Boulder IAMP Proceedings.

54. E. Mourre, Operateurs conjugués et propriétés de propagation, II, preprint.

55. G. Hagedorn, Asymptotic completeness for classes of two, three, and four particle Schrödinger operators, Trans. AMS 258 (1980), 1-75.

56. I.M. Sigal, On quantum mechanics of many-body systems with dilation-analytic potentials, Bull. AMS 84 (1978), 152-154.

57. E. Mourre and I.M. Sigal, Phase space analysis and scattering theory for N-particle systems, preprint.

58. I.M. Sigal, Mathematical foundations of quantum scattering theory for multi-particle systems, Mem. Am. Math. Soc. 209 (1978).

59. E. Mourre, Absence of singular spectrum for certain self-adjoint operators, Commun. Math. Phys. 78 (1981), 391.

60. R. Froese and I. Herbst, A new proof of the Mourre estimate, Duke Math. J. 49 (1982), 1075.

61. E. Mourre, Operateurs conjugués et propriétés de propagation, I, Commun. Math. Phys. 91 (1983), 279.

62. A. Jensen, E. Mourre and P. Perry, Multiple commutator estimates and resolvent smoothness in quantum scattering theory, preprint, 1983.

63. R. Froese and I. Herbst, Exponential bounds and absence of positive eigenvalues for N-body Schrödinger operators, Commun. Math. Phys. 87 (1982), 429.

64. R. Froese, I. Herbst, M. Hoffmann-Ostenhof and T. Hoffmann-Ostenhof, L^2-exponential lower bounds to solutions of the Schrodinger equation, Commun. Math. Phys. 87 (1982), 265-286.

65. B. Simon and B. Souillard, Franco-American meeting on the mathematics of random and almost periodic potentials, J. Stat. Phys., to appear.

66. R. Prange, D. Grempel and S. Fishman, A solvable model of quantum motion in an incommensurate potential, Phys. Rev., in press; Localization in an incommensurate potential: An exactly solvable model, Phys. Rev. Lett. 49 (1982), 833.

67. B. Simon, Almost periodic Schrödinger operators, IV. The Maryland model, Ann. Phys., to appear.

68. L. Pastur and A. Figotin, Localization in an incommensurate potential: Exactly solvable multidimensional model, JETP Lett. 37 (1983), 686; paper to appear in Commun. Math. Phys.

69. H. Kunz and B. Souillard, Sur le spectre des opérateurs aux différences finies aléatoires, Commun. Math. Phys. 78 (1980), 201-246.

70. W. Craig and B. Simon, Log Hölder continuity of the integrated density of states for stochastic Jacobi matrices, Commun. Math. Phys. 90 (1983), 207-218.

71. F. Delyon and B. Souillard, Remark on the continuity of the density of states of ergodic finite difference operators, Commun. Math. Phys., to appear.

72. F. Delyon, H. Kunz and B. Souillard, One dimensional wave equations in disordered media, J. Phys. A16 (1983), 25.

73. I. Goldsheid, S. Molchanov and L. Pastur, A pure point spectrum of the stochastic and one dimensional Schrödinger equation, Funct. Anal. Appl. 11 (1977), 1-10.

74. R. Carmona, Exponential localization in one dimensional disordered systems, Duke Math. J. 49 (1982), 191.

75. B. Simon, Some Jacobi matrices with decaying potential and dense point spectrum, Commun. Math. Phys. 87 (1982), 253-258.

76. F. Delyon, B. Simon and B. Souillard, From power law localized to extended states in a disordered system, preprint.

77. J. Moser, An example of a Schrödinger equation with an almost periodic potential and nowhere dense spectrum, Comm. Math. Helv. 56 (1981), 198.

78. V. Chulaevsky, On perturbations of a Schrödinger operator with periodic potential, Russian Math. Surveys 36, No. 5 (1981), 143.

79. J. Avron and B. Simon, Almost periodic Schrödinger operators, I. Limit periodic potentials, Commun. Math. Phys. 82 (1982), 101-120.

80. J. Avron and B. Simon, Transient and recurrent spectrum, J. Func. Anal. 43 (1981), 1-31.

81. W. Craig, Pure point spectrum for discrete almost periodic Schrödinger operators, Commun. Math. Phys. 88 (1983), 113-131.

82. J. Pöschel, Examples of discrete Schrödinger operators with pure point spectrum, Commun. Math. Phys. 88 (1983), 447-463.

83. J. Bellissard, R. Lima and E. Scoppola, Localization in ν-dimensional incommensurate structures, Commun. Math. Phys. 88 (1983), 465-477.

84. P. Sarnak, Spectral behavior of quasi periodic potentials, Commun. Math. Phys. 84 (1982), 377-401.

85. J. Avron and B. Simon, Almost periodic Schrödinger operators, II. The integrated density of states, Duke Math. J. 50 (1983), 369-391.

86. S. Aubry and G. Andre, Analyticity breaking and Anderson localization in incommensurate lattices, Ann. Israel Phys. Soc. 3 (1980), 133.

87. A. Gordon, Usp. Math. Nauk. 31 (1976), 257.

88. S. Kotani, Proceedings of the Conference on Stochastic Processes, Kyoto, 1982.

89. B. Simon, Kotani theory for one dimensional stochastic Jacobi matrices, Commun. Math. Phys. 89 (1983), 227.

STABILITY AND INSTABILITY IN QUANTUM MECHANICS

J. Bellissard

I - INTRODUCTION

The stability of the planetary system was one of the most striking problem of the nineteen century.

The perturbation theory devellopped in the late eighteen century was able to predict the motion of the celestial bodies with an amazing accuracy for time short compare to the astronomical scales. However the question of stability on the long run was not answered. In addition, the perturbative computations exhibited what was called "secular terms" namely contribution linear in time, suggesting a source of instability.

To get rid of these terms, the definition of action angle variables for nearly integrable systems, allowed the astronomers to use Fourier's expansion instead, in the perturbation theory. The price to be paid was the occurrence of small divisors, which raised the question of the convergence of the series.

The first important result toward the existence of global stability was a negative one : in 1870, Poincaré [23b] in a famous mémoire, proved that the three body problem have no more that the ten first integrals predicted by the Galilean invariance. In other word the system was not completely integrable. Quite rapidly, Poincaré realized that for nearly integrable systems periodic orbits do survive but half of them have positive Lyapounov exponent [23a,21] leading to linear instability of these orbits. It is interesting to note that in his book "Les Méthodes Nouvelles de la Mécanique Céleste" [23a], Poincaré computed these exponents almost fifteen years before the contribution of Lyapounov. He also realized how intricates must be the orbit with initial conditions close to an unstable periodic one [23a]. The contribution of Smale [30] with the "horseshoe" phenomena, and the application of the scheme to classical mechanichs by Sitnikov [22,77] gave some light on this problem.

These instabilities, which occur even at small perturbation, and must not be confused with the transition to chaos, are essentially produced in perturbation theory by the small divisors, namely the occurrence of strong resonnances. In the turn of the century the question of the convergence of the perturbation expansion was a puzzling problem which creates a long and famous discussion between Poincaré and Weïerstrass. An interesting contribution in this respect, which has been recurrently used by any person starting in the subject was the mémoire of E.Borel [7] on monogenic functions. This tool have not been powerful enough yet to provide strong results.

CPT-84/P.1659

During the first half of the twentieth century, most of the progresses concerned the description of the mathematical framework needed for a precise description of the intricated notion appearing in the stability problem. The most important contribution in this respect came from G.D. Birkhoff [5,6] especially for systems with two degrees of freedom, and for devellopping the right tools in ergodic theory. In 1939, Hopf [55] found the first example of classical system with ergodic motion : the geodisic flow on a surface of constant negative curvature. Another famous example is given by the Sinaï billiard [76] which can be consider as a limiting case of the previous example.

The stability problem required even more time. One step was done by Bogoliubov and Krylov [41] in 1934, using canonical change of variables to decrease the effective coupling. One more step was performed by C. Siegel [28, 75] during the forties, by pointing out the role of diophantine approximation in dealing with the small divisors.

It was not until 1954 however that Kolmogoroff [19,62] proposed a scheme to prove that nearly integrable systems have a lot of invariant torii. By "a lot" we mean that the union of these invariant torii is a set of positive Lebesgue measure, whose measure tends to be the full one as the amplitude of the perturbation converges toward zero. It was however only in 1962 that simultaneously Arnold [33] and J. Moser [66] gave a complete proof of this fact each in a different context, introducing the technical machinary which is nowadays considered as classical.

Since this time, the increasing use of computers allows the physicists to become much more familiar with the concrete nature of the problem. A huge litterature of theoretical works and numerical experiments on this problem was published during the seventies [20]. A scheme for describing the transition from a regular to a chaotic behaviour emerged, at least for Hamiltonian systems with two degrees of freedom, or equivalently, for area preserving maps of the cylinder. Eventhough the mathematics is far from beeing complete, a lot of progresses have been done, and the heuristics or the numerical properties of these systems is well under control.

Let us mention especially among the works on Hamiltonian systems with two degrees of freedom, the contributions of Chirikov [10] , Percival [67] , Greene [52], Doveil-Escande [12,49,50,51] , for theoretical non rigourous contributions, and S.Aubry [31,32], J. Mather [64], for mathematically rigourous contribution concerning the transition to chaos. Obviously we don't intend to give an exhaustive review of the subject, but only to emphasize on those contributions which are relevant for our subject.

The problem of quantum instabilities appeared only recently in the litterature. One of the motivations was related to the problem. In quantum chemistry of highly excited states, requiring a much better knowledge of the semi-classical region. The question was adressed to know the precise connection between classical chaotic behaviour and its quantum counter part. Attention was given to the distribution of eigenvalues, the shape of the wave functions, the semi-classical quantization procedure.

Several reviews of these problems can be found in [8, 9, 24, 26, 29] we don't want to discuss this aspect of the subject here.

Closer to the spirit of our approach is the proposal by Chirikov [10] and Taylor [78] , to look at the quantum version of the so called "standard map" :

$$q_{n+1} = q_n + p_{n+1} \qquad p_{n+1} = p_n - k \operatorname{Sin} q_n \tag{1}$$

It can be viewed as the Poincaré map for the time dependant hamiltonian system :

$$\mathcal{H}(p,q,t) = \frac{p^2}{2} + k\cos q \sum_{n\in\mathbb{Z}} \delta(t-n) \tag{2}$$

The main property of this map is the existence of a critical value $k_c = .9716...$, such that if $k \leq k_c$, there are invariant torii, whereas for $k > k_c$, no invariant torii do exist any longer. In this later case, the "typical" classical orbit wanders in the phase space (identified with $\{(q,p);\ q\in \mathbb{R}/_{2\pi\mathbb{Z}},\ p\in\mathbb{R}\} = \mathbb{T}\times\mathbb{R}$), whereas in the former case, since invariant torii do exist, the typical trajectory is trapped in bounded region separated by these torii.

After changing the period of the kicks and rescaling, the quantum analog, which is called the "kicked-Rotator model" is described by the hamiltonian

$$H(t) = -\alpha \frac{\partial^2}{\partial x^2} + k \cos x \sum_{n\in\mathbb{Z}} \delta(t-2n\pi) \tag{3}$$

acting on $L^2(\mathbb{T})$ with periodic boundary condition. As we shall see later on, the stability of the quantum motion will depend highly on the arithmetic character of α . A recent result of Casati and Guarneri [42] shows that there is a dense G_δ set of α's in $\mathbb{R}$ for which the Floquet operator corresponding to H(t) has only continuous spectrum. As a consequence the quantum average $\langle p(t)^2\rangle = \langle\varphi/p(t)^2\varphi\rangle$ of the kinetic energy in any state φ, is unbounded in time. This is one criterion which has been considered by theoretical physicists for instabilities.

However we shall concentrate on a smooth version of this model here namely:

$$H(t) = -\alpha \frac{\partial^2}{\partial x^2} + V(x,t) \tag{4}$$

where V is a 2π-periodic map in both x and t, holomorphic in a strip. We shall give a new result, namely that for a dense set of "big" Lebesgue measure of values of α, the Floquet operator for H(t) has only point spectrum, provided V is small enough. This implies in particular that any solution of the Schrödinger equation is almost periodic and that the kinetic energy stays bounded in each state for ever.

The question of stability in the large is still open even numerically [44, 48,54] . Most of the authors, following a suggestion of Prange et al [54] , choose another function v(q) instead of kcosq, which is easier to treat numerically, but which systematically corresponds to the case $k < 1/2$. We believe that for the kicked rotator problem, the quantum behaviour for k large must be qualitatively different that the quantum behaviour for k small (even though the Maryland group investigated the case for which k is close to 5 [83]).

It may be strange to speak about instabilities in quantum mechanics, since one of the spectacular result of its purpose was precisely to understand the extreme stability of atoms and molecules, namely systems with small number of degrees of freedom. However, it is well known that an electromagnetic external field may produce resonnances forcing the system to modify its internal structure, or even to blow up. The usual experimental way to do that was to send a laser beam in the optical range of frequency in order to get ionization.

But in the seventies, a surprising experimental result, not yet fully understood changed the previous scheme. It is especially illustrated by the experience of Bayfield and Koch [35,36,47,60,61] : a beam of hydrogen atoms prepared in a precise highly excited state (typically $63 \leqslant n \leqslant 69$) is flying through a microwave chamber. The wave frequencies were respectively choosen equal to 30MHz, 1.5 GHz, and 9.9 GHz. Even the last one represents about 40 % of the resonnant frequency for a single-photon excitation for the transition n = 66 to n = 67. It also represents 1 % of the photon frequency for excitation to the continuum. Nevertheless, whereas at small field amplitude the ionization rate is almost zero, there is a threshod effect at a value approximatly equal to 20 V/cm (for frequency equal to 9.9 GHz) above which the ionization rate becomes almost one. The value of the threshold depends on the frequency of the external wave, but it is observed even for the value 30 MHz, for which the physicists consider usually that a static approximation is valid.

Clearly this is a highly non linear effect which cannot be explained in term of the usual perturbation theory for it involves to compute at least up to the 100th order term. The usual tunnelling in the Stark effect does not give either a good quantitative agreement with the datas [47,60,61].

A numerical simulation was performed in 1978 by Leopold and Percival [63] using classical dynamics, and trying to take into account all source of errors. They got an excellent agreement with the experimental results of Bayfield and Koch. However no theoretical explanation was given allowing to catch a mathematical method to solve the problem. More recently R. Jensen [58,59] proposed a theoretical framework, using again the analysis of a classical system. He considered a one dimensional hydrogen atom, which is a very good model describing surface state electron on liquid helium. The corresponding hamiltonian system has one degree of freedom. Thus in a presence of a microwave field, we can use the results known for systems with two degrees of freedom. The question again is to understand large perturbation effects.

These two problems, the kicked rotator model and the Bayfield Koch experiment appear as a challenge for mathematical physicists who must develop the right framework to analyse these questions. On the long run we may be able to get rid off or to control the semi classical approximation.

In this lecture we shall concentrate on few numbers of rigourous results which are available right now. Most of them concern only the small coupling case, where both stability and instabilities are exhibit. We will organize it as follows : the section II is devoted to the question of small divisors. Basic notions in number theory are introduced. The section III is devoted to give a frame to the question of stability. In section IV we will give the known results concerning the perturbed rotator model.

ACKNOWLEDGMENTS :

The author is indebted to the Department of Mathematic of the Princeton University and especially to E. Lieb and A. Wightman, for financial and scientific support during the period when this work was initiated. He also thanks the Zentrum für Interdisciplinäre Forshung, where this work was written. He also thanks G. Casati, R. Jensen and Y. Pomeau for informations concerning the kicked rotator model or the Bayfield and Koch experiment.

II - SMALL DIVISORS

In most of the problems involving small divisors, there is one equation to be investigated a model of which is given below. For a more complete treatment and precise estimate we refer the reader to the work of Rüssmann [70,71,72,73].

The problem can be settled in the following single form :

given $g \in \mathcal{C}^{\infty}(\mathbb{T}^2)$ and $\alpha \in \mathbb{R}$, is there $f \in \mathcal{C}^{\infty}(\mathbb{T}^2)$ (or even $f \in L^2(\mathbb{T}^2)$) such that :

$$\frac{\partial f}{\partial x} + \alpha \frac{\partial f}{\partial y} = g(x,y) \tag{1}$$

here we use the convention $\mathbb{T} = \mathbb{R}/_{2\pi\mathbb{Z}}$.

1) Formal solutions

To solve (1) we first compute the Fourier transform of both side. If

$$f(x,y) = \sum_{(m,n)\in\mathbb{Z}^2} \hat{f}(m,n) \exp i\,(mx + ny) \tag{2}$$

we get :

$$i\,(m + n\alpha)\hat{f}(m,n) = \hat{g}(m,n) \tag{3}$$

To get a solution of (3) it is necessary and sufficient that

$$m + n\alpha = o \Rightarrow \hat{g}(m,n) = o \tag{4}$$

We must consider at this point two cases :

Case a : $\alpha \in \mathbb{Q}$

If $\alpha = p/_q$ is rationnal, where p,q are prime to each other, the condition (4) means

$$\hat{g}\,(-lp, lq) = o \qquad \forall l \in \mathbb{Z} \tag{5}$$

Equivalently we get :

$$\int_0^{2\pi q} ds\; g(x + s,\; y + sp/_q) = o \tag{6}$$

If the condition (5)(6) is satisfied, then

$$mq + np \neq o \qquad \hat{f}(m,n) = q\,\frac{\hat{g}(m,n)}{i(mq+np)} \tag{7}$$

Since mq + np is a non zero integer, and $g \in \mathcal{C}^{\infty}(\mathbb{T}^2)$, f must be a fast decreasing sequence, and therefore $f \in \mathcal{C}^{\infty}(\mathbb{T}^2)$ exists.

Proposition 1. If $\alpha = p/_q$ is rationnal the equation (1) has a $\mathcal{C}^{\infty}$ solution if and only if

$$\int_0^{2\pi q} ds\; g(x + s,\, y + \frac{sp}{q}) = o \text{ for all } (x,y) \in \mathbb{T}^2$$

If this condition is not satisfied, there is no formal solution.

The only if part can be immediatly checked from (1) by integrating both side of the path $s \longmapsto (x + s,\, y + \frac{sp}{q})$ $(s \in [o, 2\pi q])$.

Case b : $\alpha \notin \mathbb{Q}$

If α is irrationnal the condition (4) is equivalent to :

$$o = \hat{g}(o,o) = \int_{\mathbb{T}^2} g(x,y)\,dxdy \equiv \langle g \rangle \tag{8}$$

This condition is obviously necessary; it is also sufficient to get a formal solution :

$$(m,n) \neq o \Rightarrow \hat{f}(m,n) = \frac{\hat{g}(m,n)}{i(m + \alpha n)} \tag{9}$$

The next question is to know wether or not (9) allows us to get f in $\mathcal{C}^{\infty}(\mathbb{T}^2)$ or even in $L^2(\mathbb{T}^2)$. Because α is irrationnal, it can be approximated by rationnal, and for a certain subsequence of pairs (m,n), the divisor in the right hand side of (9) can be fairly small. Sometimes it may be so small that the Fourier series corresponding to (9) will diverge.

2) Elements of number theory :

Hurwitz's theorem [15,18,27] : if α is an irrationnal number, there is a sequence $(p_n/q_n)_{n\in\mathbb{N}}$ of rationnal numbers such that :

$$|\alpha - p_n/q_n| \leq \frac{1}{\sqrt{5}}\frac{1}{q_n^2}$$

This bound is saturated for the golden mean $\alpha = \frac{\sqrt{5}-1}{2}$

Thanks to this result, $|q_n\alpha - p_n|^{-1}$ diverges at least like q_n. In order to get a L^2 solution from (9) it is therefore necessary that

$$\limsup_{n\to\infty} q_n \cdot \hat{g}(-p_n,q_n) = 0$$

However this is far from being sufficient.

Definition (i) An irrationnal number α is called diophantine of power σ if there is $\gamma > 0$ such that

$$|\alpha - p/q| \geq \gamma/q^{\sigma} \qquad \forall\, p/q \in \mathbb{Q}$$

(ii) α is a Liouville number if it is neither rational nor diophantine. Equivalently it exists a sequence $(p_n/q_n)_{n\in\mathbb{N}}$ of rationnal such that

$$|\alpha - p_n/q_n| < (q_n)^{-n}$$

We first remark that from a theorem of Liouville [27] every algebraic number is diophantine.

On the other hand the number

$$\alpha = \sum_{n=1}^{\infty} 2^{-n!}$$

is certainly a Liouville number. Thus non algebraic numbers do exist. This was the first proof of existence of transcendant numbers.

Improvements of the Liouville theorem lasted for more than a century. Contribution by Thue [79] in 1909, Siegel [74] in 1921, and finally by Roth [69] in 1955 gave the result that any algebraic number is diophantine of power $2+\varepsilon$ (any $\varepsilon > 0$).

Actually probabilistic and topological methods give the :

Theorem 2 i) The set of diophantine numbers of power σ has full Lebesgue measure if and only if $\sigma > 2$.
ii) The set of Liouville numbers is a dense G_δ subset of $\mathbb{R}$, with zero Lebesgue measure.

Let us recall that a G_δ set is a countable intersection of open set. By the Baire theorem, a countable intersection of dense open set is a dense G_δ. A generic property is a property true on a dense G_δ. Thus being a Liouville number is a generic property even though from measure theoretic point of view, they are extremely rare. Here topology and measure theory disagree.

3) Formal and global solutions :

Theorem 3 If α is diophantine, and if $g \in \mathcal{C}^\infty(\mathbb{T}^2)$ satisfies (8), the equation (1) has always a $\mathcal{C}^\infty$ solution. Two such solutions differ by a constant.

The proof is obvious from (9) and the definition of diophantine numbers.

This theorem can be supplemented by a more precise result, which tells us where does the small divisor produce a result : this is the question of the loss of smoothness.

Let g be a holomorphic function in the strip $B_r = \{(x,y) \in \mathbb{T}^2 + i\mathbb{R}^2;$ $|\mathrm{Im}\, x| < r,\ |\mathrm{Im}\, y| < r$ with the condition :

$$\|g\|_r = \sup_{(x,y) \in B_r} |g(x,y)|$$

We get :

Theorem 4 (Rüssmann [72]) If $\|g\|_r < +\infty$ and satisfies (8), the solution given by (9) defines a holomorphic solution of (1) in B_r which is not bounded in B_r. Moreover there is a constant depending only on σ, such that if

$$|\alpha - p/q| \geq \gamma / q^\sigma \qquad \forall\, p/q \in \mathbb{Q}$$

then :

$$\|f\|_{r-\rho} \leq \frac{C(\sigma)}{\gamma \rho^\sigma} \|g\|_r. \tag{10}$$

This estimates would not be difficult to get from the condition on α, if the norm $\|\cdot\|_r$ were replaced by

$$|||f|||_r = \sum_{(m,n)\in\mathbb{Z}^2} \hat{f}(m,n)\, e^{r(|m|+|n|)}$$

However, the necessity of using twice the Fourier transform make this estimate difficult. It requires a subtle analysis of the small divisor.

Another consequence of (10) is that it is possible to solve (1) when g is only k-times differentiable, using a method introduced by J. Moser. Then the solution f is in $\mathcal{C}^{k-\sigma}$ in general.

In the Liouville case the situation is more involved. We shall give two kinds of results depending on the problem.

<u>Theorem</u> 5 [see 17] If α is a Liouville number, there is $g \in \mathcal{C}^\infty(\mathbb{T}^2)$ satisfying(8), such that (1) has no distribution solutions.

The proof of this theorem is direct using the Fourier transform. If f is a distribution then $\{\hat{f}(m,n); (m,n)\in\mathbb{Z}^2\}$ is polynomially bounded. Choosing α Liouville and $\hat{g}$ fast decreasing but not too fast, it is always possible to get $\hat{f}$ not polynomially bounded.

The previous result is actually a property of the operator

$$L_\alpha(f) = \frac{\partial f}{\partial x} + \alpha \frac{\partial f}{\partial y}$$

Theorem 5 says that L_α is not onto $\mathcal{C}^\infty(\mathbb{T}^2)$. However, if g is given, an another result is useful.

<u>Theorem</u> 6 For a generic g in $\mathcal{C}^\infty(\mathbb{T}^2)$, there is a dense G_δ set $\Omega(g)$ in $\mathbb{R}$, such that if $\alpha\in\Omega(g)$, the equation :

$$\frac{\partial f}{\partial x} + \alpha \frac{\partial f}{\partial y} = g - \langle g\rangle \qquad (11)$$

has no L^2 solution.

We shall give the proof of this result, for it is a typical one, which has been used by J. Avron and B. Simon [34]for the almost Mathieu equation, and more recently by G. Casati and Guarneri [43] for the kicked rotator model. The idea consists in considering the cas $\alpha = p/_q$, and getting precise estimates; then to compare α with rationnals.

Proof :

Without loss of generality we may assume $\langle g \rangle = o$. Then it is easy to see that (11) is equivalent to

$$f(x+t,y+t\alpha) - f(x,y) = \int_0^t ds\, g(x+s,y+\alpha s) \equiv \Gamma_{\alpha,t}(x,y) \tag{12}$$

We first evaluate the variation of the right hand side as α varies.

Lemma 1 :

We have

$$\| \Gamma_{\alpha,t} - \Gamma_{\alpha',t} \|_{L^2} \leq t^2/\sqrt{3} \; |\alpha-\alpha'| \; \| \nabla g \|_{L^2} \tag{13}$$

Proof of Lemma 1 : We get :

$$\int dxdy \left| \Gamma_{\alpha,t} - \Gamma_{\alpha',t} \right|^2 \leq \int dxdy \left| \int_0^t ds \int_0^1 d\sigma \frac{\partial g}{\partial y}(x+s, y+[\alpha\sigma + (1-\sigma)\alpha'] s) \right|^2 |\alpha - \alpha'|^2$$

Using the Schwarz inequality, and the Fubini theorem, we get :

$$\leq \frac{t^3}{3} |\alpha - \alpha'|^2 \int_0^t ds \int_0^1 d\sigma \int dxdy \left| \frac{\partial g}{\partial y}(x+s, y+(\sigma\alpha + (1-\sigma)\alpha')s) \right|^2$$

Changing variables in the (x,y) integral, we get :

$$\leq \frac{t^3}{3} |\alpha - \alpha'|^2 \int_0^t ds \int_0^1 d\sigma \; \| \nabla g \|_{L^2}^2 = \frac{t^4}{3} |\alpha - \alpha'|^2 \| \nabla g \|_{L^2}^2$$

The next step consists in evaluating the time behaviour of $\Gamma_{\alpha,t}$ when α is rationnal.

Lemma 2 :

There are two functions Φ, F on $\mathbb{T}^2$ such that :

$$\Gamma_{\frac{p}{q},t}(x,y) = t\Phi(x,y) + q\left[F(x+t,y+t\tfrac{p}{q}) - F(x,y)\right] \qquad (14a)$$

where :

$$\Phi(x,y) = \sum_{l\neq o} \hat{g}(-lp,lq)\,e^{il(yq-px)} = \int_{o}^{2\pi q} \frac{ds}{2\pi q}\, g(x+s,y+s\tfrac{p}{q}) \qquad (14b)$$

and

$$F(x,y) = \sum_{\substack{(m,n)\in\mathbb{Z}^2 \\ mq+np\neq o}} \frac{\hat{g}(m,n)}{{}^{2}(mq+np)}\, e^{i(mx+ny)} \qquad (14c)$$

In addition

$$\|F\|_{L^2} \leq \|g\|_{L^2} \qquad (14d)$$

Proof :

Immediate from Fourier's expansion of $\Gamma_{p/q\,t}$, and the Plancherel formula.

Remark :

The first term $t\Phi(x,y)$ corresponds exactly to the secular term of the celestial mechanics.

Let us consider the subset Σ of $\{g\in\mathcal{C}^{\infty}(\mathbb{T}^2);\ \langle g\rangle = o\}$ the elements of which have no vanishing Fourier coefficients. Clearly Σ is a countable intersection of dense open sets. Consider $g\in\Sigma$, and let us define (using Plancherel's formula)

$$\eta_g(p,q) = \Big(\sum_{l\neq o} \hat{g}(-lp,lq)^{\ 2}\Big)^{1/2} = \|\Phi\|_{L^2} \qquad (15)$$

By hypothesis, $\eta_g(p,q) > o$. Since $g\in\mathcal{C}^{\infty}$, for each $N\in\mathbb{N}$, there is $C_N > o$ such that :

$$\eta_g(p,q) \leq \frac{C_N}{(|p|+|q|)^N} \qquad (16)$$

From eq. (14a) (14d) in lemma 2, we get : from the triangle inequality

$$\|\Gamma_{p/q\,t}\|_{L^2} \geq t\,\eta_g(p,q) - 2q\,\|g\|_{L^2} \qquad (17)$$

Using (13), for any $p/_q \in \mathbb{Q}$, we get :

$$\|\Gamma_{\alpha,t}\|_{L^2} \geq \|\Gamma_{p/_q,t}\|_{L^2} - \|\Gamma_{\alpha,t} - \Gamma_{p/_q t}\|_{L^2}$$

Which gives :

$$\|\Gamma_{\alpha t}\|_{L^2} \geq t\,\eta_g(p,q) - 2q\|g\|_{L^2} - \frac{t^2}{\sqrt{3}}|\alpha - p/_q|\,\|\nabla g\|_{L^2} \tag{18}$$

Let us now define $\Omega(g)$ as follows :

$$\Omega(g) = \bigcap_{N\in\mathbb{N}_*} \left\{\alpha\in\mathbb{R} \; ; \; \exists\, p/_q \in \mathbb{Q}\, |\alpha - \frac{p}{q}| < \frac{1}{N}\frac{\eta_g(p,q)^2}{q^2}\right\} \tag{19}$$

Clearly $\Omega(g)$ is the intersection of a countable family of open sets, each of which containing $\mathbb{Q}$. Thus $\Omega(g)$ is a dense G_δ. It is not difficult, using (16) to see that each of these open sets have a Lebesgue measure dominated by $\text{const}/_N$. Therefore $\Omega(g)$ has zero Lebesgue measure.

In addition, if $\alpha \in \Omega(g)$, for each $N \in \mathbb{N}_*$ there is $p_N/_{q_N} \in \mathbb{Q}$, such that

$$|\alpha - p_N/_{q_N}| < \frac{1}{N}\frac{\eta_g(p_N,q_N)^2}{{q_N}^2}$$

If α is irrationnal (which is possible since $\Omega(g)$ cannot be countable), $q_N \to \infty$ as $N \to \infty$. Let us choose

$$t_N = N^{1/2} q_N\, \eta_g(p_N,q_N)^{-1}$$

Thanks to (16), $t_N \to \infty$, and from (18) we get :

$$\|\Gamma_{\alpha,t_N}\|_{L^2} \geq (N^{1/2} - 2\|g\|_{L^2})q_N - \frac{\|\nabla g\|_{L^2}}{\sqrt{3}} \tag{20}$$

Consequently, $\limsup_{t\to\infty} \|\Gamma_{\alpha,t}\|_{L^2}$ is infinite for $\alpha\in\Omega(g)$, which contradicts the existence of f in L^2 satisfying (12). For if f exists we must have :

$$2\|f\|_{L^2} \geq \|\Gamma_{\alpha,t}\|_{L^2} \quad \forall t. \in \mathbb{R}\,.$$

III - TIME DEPENDENT HAMILTONIAN [see 56,80]

We want to concentrate of the following type of Schrödinger operators :

$$H(t) = H_o + V(t) \tag{1}$$

Where H_o is a positive self adjoint unbounded operator whith discrete spectrum, on a Hilbert space $\mathcal{H}$. We will choose $V(t)$ as a norm continuous one parameter family of bounded self adjoint operators, whith the additionnal property

$$V(t + 2\pi) = V(t)$$

Classical examples of operators H_o are given by elliptic pseudodifferential operators of positive degree, on a compact manifold. One of the simplest example being $H_o = -\frac{\partial^2}{\partial x^2}$ on $\mathbb{T}$. In other words we are looking at the quantum version of a classical system with compact configuration space, periodically perturbed by a time dependant potential.

The Schrödinger equation

$$i\frac{\partial\psi}{\partial t} = H(t)\psi \tag{2}$$

Can be solved via an evolution operator $\mathbf{U}(t,s)$ which can be constructed from a Dyson expansion after performing the "interaction picture" in (2). As a result $U(t,s)$ satisfies the following properties :

i) it is unitary $U(t,s)^* = U(t,s)^{-1} = U(s,t)$. $s,t \in \mathbb{R}$

ii) it is an evolution : $U(t,s) = U(t,u)U(u,s)$ $s,t,u \in \mathbb{R}$

iii) it is periodic $U(t+2\pi, s+2\pi) = U(t,s)$, $s,t \in \mathbb{R}$

iv) it is strongly continuous in both s and t .

Thanks to these properties it is sufficient to consider the Floquet operator defined by

$$U_V = U(2\pi, o) \tag{3}$$

to describe the long time behaviour of the wave function. For $V = o$, $U_o = \exp 2i\pi H_o$

has pure point spectrum. This spectrum is either made of isolated eigenvalues in which cases they may be infinitely degenerate , or there are limit points, in which case we may get dense point spectrum.

To express the stability question, let us consider a state $\varphi \in \mathcal{H}$, $\|\varphi\|=1$, and the time evolution of the kinetic energy :

$$\mathcal{E}_{\varphi}(t) = \langle \varphi | U_v^{\,t} H_0 U_v^{\,-t} \varphi \rangle, \quad t \in \mathbb{Z} \tag{4}$$

We get the following :

<u>Theorem</u> 7 [82] If φ belongs to the continuous spectral subspace of U_v, then $\mathcal{E}_{\varphi}$ is unbounded in time.

<u>Proof</u> :

Let $\{e_n\}_{n\geq o}$ be an orthonormal basis of eigenvectors of Ho, labelled in such a way that the corresponding eigenvalues $E_n = \langle e_n | H_o e_n \rangle$ are ordered :

$$E_0 \leq E_1 \leq E_2 \leq \ldots .$$

Since H_o is unbounded, $E_n \to \infty$ as $n \to \infty$. Since the spectrum is discrete, the multiplicity of each eigenvalue is finite, and there is no limit point at finite distance. For $\varphi \in \mathcal{H}_{cont}(U_v)$, $\|\varphi\|=1$ (the continuous spectral subspace of U_v) we set

$$C_n(t) = |\langle e_n | U_v^{\,t} \varphi \rangle|^2 \tag{5}$$

Then we get clearly :

$$\sum_{n=o}^{\infty} C_n(t) = \|\varphi\|^2 = 1 \tag{6}$$

On the other hand, by the Wiener criteria, or the RAGE theorem [25] we get for each n :

$$\lim_{T\to\infty} \frac{1}{T} \sum_{t=o}^{T-1} C_n(t) = o$$

Therefore given $\varepsilon > o$, and $N \in \mathbb{N}$, there is $T(\varepsilon, N) \in \mathbb{N}_*$ such that :

$$\frac{1}{T} \sum_{t=o}^{T-1} \sum_{n=o}^{N} C_n(t) \leq \varepsilon \quad \forall T \geq T(\varepsilon, N) \tag{7}$$

By (6) we get : for $T \geq T(\varepsilon, N)$

$$\frac{1}{T}\sum_{t=o}^{T-1}\sum_{n=N+1}^{\infty} C_n(t) \geq 1-\varepsilon$$

Therefore : for $T \geq T(\varepsilon, N)$

$$\frac{1}{T}\sum_{t=o}^{T-1} \mathcal{E}_{\varphi}(t) \geq \frac{1}{T}\sum_{t=o}^{T-1}\sum_{n=N+1}^{\infty} E_n C_n(t) \geq E_{N+1}(1-\varepsilon)$$

Since $\lim_{N\to\infty} E_{N+1} = \infty$, $\mathcal{E}_{\varphi}(t)$ cannot be bounded.

As a consequence, stability will occur globally only if U_v has only point spectrum. In other words global stability implies that all solution of the Schrödinger equation must be almost periodic, as a function of the time.

This is a surprising result if we compare to the case of classical mechanics, at least for systems with two degrees of freedom. In this later situation, the K.A.M theorem insures that global stability does occur since the phase space is disconnected by invariant torii. However, not all orbits are almost periodic, since there are hyperbolic periodic orbits.

In order to investigate spectral properties of U_v, there is a method, avoiding the computation of the spectrum of a unitary operator, replacing it by a self adjoint one. It is a well known method in classical mechanics, which consists in replacing a time dependant hamiltonian with N degrees of freedom by a time independant one having (N+1) degrees of freedom, giving rise to the same equations of motion. It has been used in quantum mechanics by Howland [56] and Yajima [80] .

Let us consider the operator

$$K_v = -i\frac{\partial}{\partial t} + H_o + V(t)$$

acting on $\mathcal{K} = L^2(\mathbb{T})\otimes\mathfrak{h}$ namely the hilbert space of measurable functions $\mathbb{R}\ni t\mapsto f(t)\in\mathfrak{h}$ whith $f(t+2\pi) = f(t)$ and $\int_0^{2\pi}\|f(t)\|^2_{\mathfrak{h}}\frac{dt}{2\pi} = \|f\|^2_{\mathcal{K}} < +\infty$. There is a domain on which K_v becomes a self adjoint operator.

The advantage of considering K_v instead of U_v is described in the following :

Yajima's Theorem [80] : There is a unitary operator W on $\mathcal{K}$, such that

$$\exp 2i\pi K_V = W(\mathbb{1} \otimes U)W^*$$

In other words, up to multiplicity, the spectrum of U_V and $\exp 2i\pi K_V$ agree.

We shall use K_V to investigate the stability of the pulsed rotator in the next section.

IV - THE PULSED AND THE KICKED ROTATORS

Let us now come to the models shortly described in our introduction (§ I eq. 3 and 4), proposed by Chirikov et al [44,45,57] , as a good quantum model for investigating the transition to chaos.

The kicked rotator is described as an hamiltonian acting on $L^2(\mathbb{T})$, and formally given by :

$$H(t) = -\alpha \frac{\partial^2}{\partial x^2} + v(x) \sum_{n=-\infty}^{+\infty} \delta(t-2n\pi) \tag{1}$$

The δ-function may be a source of difficulties for the solutions of the Schrödinger equations given by (1) have discontinuities at $t=2n\pi$. To overcome this difficulty, let us replace δ by $\delta_\varepsilon(t)=\frac{1}{\varepsilon}\rho(\frac{t}{\varepsilon})$ where ρ is a positive $\mathcal{C}^\infty$ function with support in $[-1,+1]$, and $\int dx\rho(x) = 1$. Then there is no difficulty using an interaction scheme to compute the Floquet operator via a Dyson expansion. Since in this expansion, δ_ε appears always as integrand in integrals over time we get the following result :

Proposition 1. If $U_{\varepsilon v}$ is the Floquet operator defined by (1) with δ replaced by δ_ε, then :

$$\operatorname*{s-lim}_{\varepsilon \to 0} U_{\varepsilon, v} = \exp\left(2i\pi\alpha \frac{\partial^2}{\partial x^2}\right) \cdot \exp\left(-2i\pi v(x)\right) = U_V \tag{2}$$

Let us define the kicked rotator as an evolution whose Floquet operator is U_V.

For v=o, U_o has a pure point spectrum. The eigenvalues are the elements of

$\sigma_\alpha = \{\exp - 2i\pi\alpha n^2;\ n \in \mathbb{Z}\}$. By a result of H. Weyl, σ_α is either finite when is rationnal, or dense in S_1, if α is irrationnal. [see 11,13]. For $v \neq o$ we get actually the following result :

Proposition 2 [57]. There is a dense G_δ set Σ in $\mathcal{C}^\infty(\mathbb{T})$ such that if $v \in \Sigma$, and $\alpha = p/_q$ is rationnal, U_v has only an absolutely continuous spectrum. If $\|v\|_\infty$ is small enough, the spectrum of U_v has $N(p/_q)$ gaps where $N(p/_q)$ is the number of elements of $\sigma_{p/q}$.

Proposition 3. If α is irrationnal, the spectrum of U_v is the full circle.

Proof :

Consider the operator W_m, of multiplication by $\exp i^\circ mx$ in $L^2(\mathbb{T})$. For $(\xi,\eta) \in \mathbb{R}^2$ we define

$$U_v(\xi,\eta) = e^{2i\pi\left[\alpha\frac{\partial^2}{\partial x^2} - \frac{\xi}{i}\frac{\partial}{\partial x} - \eta\right]} e^{-2i\pi v(x)} \tag{3}$$

Clearly the translation $\eta \to \eta + 1$ does not change $U_v(\xi,\eta)$, and since $\frac{1}{i}\frac{\partial}{\partial x}$ admits $\mathbb{Z}$ as spectrum, the translation $\xi \to \xi + 1$ does not change $U_v(\xi,\eta)$ either. Thus (ξ,η) can be consider as elements of $(\mathbb{R}/_\mathbb{Z})^2$.

Now we get easily :

$$W_m U_v(\xi,\eta) W_m^* = U_v(\xi + 2m\alpha, \eta + m\xi + \alpha m^2) \tag{4}$$

On the other hand, $(\xi,\eta) \mapsto U_v(\xi,\eta)$ is strongly continuous. Now, by H. Weyl's result [11,13], given η_o in $\mathbb{R}/_\mathbb{Z}$ it is possible to find a sequence $(m_k)_{k\in\mathbb{N}}$ of integers such that

$$\lim_{k\to\infty} (m_k\alpha) = o \pmod 1 \qquad \lim_{k\to\infty} (\alpha\, m_k^2) = \eta_o \pmod 1$$

Thus remarking that $U_v = U_v\,(o,o)$ we get :

$$\text{s-}\lim_{k\to\infty} W_{m_k} U_v(o,o) W_{m_k}^* = U_v(o,\eta_o) = e^{-2i\pi\eta_o}\, U_v(o,o) \tag{5}$$

By strong limit the spectrum does not increase [25], and since U_v is unitary we get :

$$e^{-2i\pi\eta_0}\,\sigma(U_v) \subset \sigma(U_v) \subset S_1$$

Since η_0 is arbitrary, if $z \in \sigma(U_v)$, $\sigma(U_v) \supset \{e^{-2i\pi\eta_0} z \ ; \eta_0 \in \mathbb{R}/\mathbb{Z}\} = S_1$

One important contribution in the study of this subject was made by the Maryland group, namely Fishman, Grempel and Prange [53,54]. They actually proved :

<u>Proposition</u> 4 Assume that $\sup_{x \in \mathbb{R}} |v(x)| < 1/2$. Then the eigenvalue equation

$$U_v \psi = e^{-2i\pi\omega} \psi$$

is equivalent to the equation

$$\tan \pi(n^2\alpha - \omega)\, \hat{\varphi}(n) + \sum_{n' \in \mathbb{Z}} \widehat{\tan \pi v}\,(n-n')\, \hat{\varphi}(n') = 0 \tag{6}$$

where $(\hat{\varphi}(n))_{n \in \mathbb{Z}}$ is the Fourier transform of the function $(1+e^{-2i\pi v})\psi$.

In particular, assume that v is holomorphe in the strip $B_r = \{x \in \mathbb{T} + i\mathbb{R} \ ; |\mathrm{Im}x| < r\}$. with $\|v\|_r < 1/2$. Then $\tan \pi v$ is holomorphic bounded in Br too, and its Fourier coefficients satisfy :

$$|\widehat{\tan \pi v}(n)| \leq \|\tan \pi v\|_r \, e^{-r|n|}, n \in \mathbb{Z} \tag{7}$$

(6) is a Schrödinger - like equation on the discrete line. The argument given by Prange et al, was that the sequence $\{\alpha n^2 \bmod 1; \ n \in \mathbb{Z}\}$, is pseudo-random as α is irrationnal. Therefore the potential

$$V_n = \tan \pi(\alpha n^2 - \omega)$$

are random variables with "weak correlation". (Actually the four point correlation is not so weak). Applying the intuition of Anderson for random potentials they suggested that (6) should have only exponentially localized solutions. This in turn would prove that U_v has only point spectrum, a stability result.

This argument was supplemented by numerical calculations, also performed by Pomeau-Dorrizzi, and Grammaticos [48]. There is a strong evidence indeed that the conclusion of this argument should be true.

However, there are certain limitation to it.

The first one is that the authors choosed α very irrationnal. The golden mean $\alpha = \frac{\sqrt{5}-1}{2}$ was the best candidate or at least a diophantine number.

The second limitation is that no prediction is done for $\|v\|_\infty \geq 1/2$. In this case (6) is still correct provided we take some precaution for $\tan \pi v$ is no longer a bounded operator, and φ must belong to its domain. The important property is that (7) fails to be true; the singularity in $\tan \pi v(x)$ coming from values such that $v(x_o) = 1/2$, produces a long range interaction. We suggest that such a long tail produce a continuous spectrum for U_v.

The answer to the first objection is given by the following important result :

<u>Theorem</u> 9 (Casati-Guarneri [43]). There is a generic set Σ in $\mathcal{C}^\infty(\mathbb{T})$ such that if $v \in \Sigma$, there is a dense G_δ set $\Omega(v)$ in $\mathbb{R}$, for which if $\alpha \in \Omega(v)$, U_v has only a continuous spectrum.

The proof of this result follows the same line as the proof we gave in section 2, Theorem 6. We must evaluate the long time behaviour of (if $\varphi \in L^2(\mathbb{T})$, $\|\varphi\| = 1$).

$$S(T,\alpha) = \frac{1}{T} \sum_{t=o}^{T-1} |\langle \varphi | U_v^t \varphi \rangle|^2 \tag{8}$$

First the asymptotics is computed quite accurately for $\alpha = p/_q$, and then, we evaluate the difference $S(T,\alpha) - S(T,\alpha')$. A subsequence $T_N \to \infty$ can be found such that $S(T_N,\alpha) \to o$ if α belongs to $\Omega(v)$; by Wiener's criterium, or the RAGE theorem [25] , we conclude that pure point spectrum is absent. Some care must be taken with the uniformity of the convergence with respect to the state.

This result shows that the argument of the Maryland group requires some subtle consideration if we want to prove it. The same kind of difficulties appeared in the study of the Almost Mathieu (or Harper) equation, where the Aubry conjecture depends on the arithmetic nature of the frequency ratio (see [29]). Actually, it turned out that the stability result was by far more difficult to prove than the instability one, for it requires the use of the K.A.M theorem. In our case here, we have not been able to produce a proof of the stability for the kicked rotator when α is sufficiently diophantine.

However if one considers the pulsed rotator, namely

$$H(t) = -\alpha \frac{\partial^2}{\partial x^2} + V(x,t) \tag{9}$$

where now V is 2π-periodic in x and t, and holomorphic in these variables in the strip B_r, whith $\|V\|_r < +\infty$, then, it is possible to conclude :

Theorem 10 [see [81]]. Assume that U_V is the Floquet operator for the hamiltonian (9). Then given $\varepsilon > 0$ and $r_\infty \in (0,r)$ there is $\mu_c > 0$, and a subset Ω of $[1,\infty)$ such that :

(i) Ω is closed, and its complement has a Lebesgue measure smaller than ε

(ii) if $\|V\|_r < \mu_c$, $\alpha \in \Omega$, U_V has only point spectrum with a dense set of eigenvalues.

(iii) the eigenvalues are given by $\{\exp - 2i\pi(\alpha m^2 + g_V(\alpha, m))\,;\ m \in \mathbb{Z}\}$ where $g_r(\alpha, m)$ is a sequence of $\mathcal{C}^1$ functions of α defined on Ω ; in addition

$$\sup_{m \in \mathbb{Z}} \| g_V(\cdot,m)\|_{\mathcal{C}^1} = \| g_V\|_{\mathcal{C}^1} ,$$

converges to zero as $\|V\|_r \to 0$.

(iv) the corresponding eigenfunction φ_m are holomorphic in B_{r_∞} and close to the unperturbed one, namely;

$$\lim_{\|V\|_r \to 0} \sup_m \| \varphi_m - e^{imx} \|_{B_r} = 0$$

The proof of this theorem will appear elsewhere [81] . It requires the use of the K.A.M method, with a detailed analysis for the construction of the set Ω. The method however has already been used for a certain class of almost periodic potential (see 38,39,46,68). In this latter case, the estimates are simpler. The reader should look at them before.

CONCLUSIONS

Many problems are still open on investigation of models to understand the instability in quantum mechanics. This lecture was not supposed to exhibit a full set of exhaustive results. It should be a starting point to convince the experts and the beginners to investigate these questions, which are possible to control nowaday, even though very delicate.

GENERAL MONOGRAPHIES

1 R. ABRAHAM, J.E. MARSDEN, Foundations of Mechanics, Benj. Reading, Mass. 2nd Ed. (1978).

2 V.I. ARNOLD, A. AVEZ, Problèmes ergodiques en Mécanique Classique, Paris (1967).

3 V.I. ARNOLD, Chapitres Supplémentaires de la théorie des équations différentielles ordinaires, Ed. Mir, Moscou (1980).

4 P. BILLINGSLEY, Ergodic theory and Information, Chap. 1, § 4, J. Wiley & Sons, New York, London, Sidney (1965).

5 G.D. BIRKHOFF, Dynamical Systems, AMS Coll. Publications, Vol. 9, (1927), Reprinted 1966.

6 G.D. BIRKHOFF, Collected Mathematical Works, Vol. 2, Dover, New York (1968).

7 E. BOREL, Leçons sur les Fonctions Monogènes uniformes d'une Variable Complexe, Gauthiers-Villars, Paris (1917).

8 G. CASATI, J. FORD ed., Stochastic Behaviour in Classical and Quantum Hamiltonian systems, Springer, Berlin, Heidelberg, New York, Lecture Notes in Physics, 93 (1979).

9 G. CASATI ed., Chaotic Behaviour in Quantum Systems, Plenum Press, New York (1984).

10 B.V. CHIRIKOV, A Universal Instability of Many Dimensional Oscillator systems, Phys. Rep., 52, 263 (1979).

11 I.P. CORNFELD, S.V. FORMIN, Ya.G. SINAI, Ergodic theory, Grundlerhren Bd 245, (1982), Springer Verlag, Berlin, Heidelberg, New York.

12 F. DOVEIL, Groupe de Renormalisation pour les Systèmes Hamiltoniens non intégrables : application au mouvement d'une particule chargée dans un paquet d'ondes longitudinal, Thèse d'Etat, Univ. Paris VI, (1982), Paris.

13 H. FURSTENBERG, Recurrence in Ergodic Theory and Combinatorial Number Theory, Princeton Univ. Press, Princeton, N.J., (1981).

14 G. GALLAVOTTI, Elements of classical Mechanics, Springer Text and Monograph in Physics (1983).

15 G.H. HARDY, E.M. WRIGHT, An Introduction to the Theory of Numbers, 5th Ed., Oxford Clarendon Press, (1979).

16 M.R. HERMAN, Sur la Conjugaison différentiable des diffeomorphismes du cercle à des rotations, Pub. de l'I.H.E.S., 49 (1979).

17 M.R. HERMAN, Sur les Courbes invariantes par les diffeomorphismes de l'anneau, Vol. 1, Asterisque 103-104 (1983).

18 A.Ya. KHINTCHINE, Continuous fractions, Chicago Univ. Press, (1964), Chicago.

19 A.N. KOLMOGOROV, Théorie générale des Systèmes dynamiques en Mécanique classique, Proc. Intern. Congress of Math., Amsterdam (1957), English translation in App. D of ref. [1].

20 A.J. LICHTENBERG, M.A. LIEBERMANN, Regular and Stochastic Motion, Springer Verlag, Berlin, Heidelberg, New York, (1983).

21 A.M. LYAPOUNOV, Problème Général de la Stabilité des Mouvements, Ann. Fac. Sci. Toulouse (2), 203-474 (1907), Reprinted in Ann. Math. Studies n° 17, (1947), Princeton Univ. Press, Princeton, N.J.

22 J. MOSER, Stable and Random Motions in Dynamical Systems, Ann. Math. Studies, (1973), Princeton Univ. Press, Princeton, N.J.

23 H. POINCARE, a) Les Méthodes Nouvelles de la Mécanique Céleste, Tomes I,II,III, Gauthiers-Villars (1892-1894-1899), Reprinted by Dover New York (1957), b) Oeuvres.

24 Proceedings of the 1981 Les Houches Summer School, Chaotic Behaviour in deterministic Systems, G. Iooss, R.H.G. Helleman, R. Stora Ed., North Holland Amsterdam (1984).

25 M. REED, B. SIMON, Method of Modern Mathematical Physics, Vol. III : Scattering Theory, Academic Press (1978), London, New York.

26 W.P. REINHARDT, Classical Chaos, The Geometry of Phase Space and Semi-Classical Quantization, To appear in The Mathematical Analysis of Physical Systems, R. Mickens Ed., Van Nostrand-Reinhold, New York, In Press.

27 W. SCHMIDT, Diophantine Approximations, Lecture Notes In Math. n° 785 (1980), Springer Verlag, Berlin, Heidelberg, New York.

28 C.L. SIEGEL, J.K. MOSER, Lectures on Celestial Mechanics, Grundlehren Bd 187, (1971), Springer Verlag, Berlin, Heidelberg, New York.

29 B. SIMON, Almost Periodic Operators, Adv. Appl. Math, 3 (1982) 463-490.

30 S. SMALE, Differentiable Dynamical Systems, Bull. AMS, 73, (1967), 747-817.

REFERENCES

31 a) S. AUBRY, P.Y. LE DAERON, The Discrete Frenkel-Kontorova model and its applications, I, Exact results for the ground States, Physica 8D (1983) 381-422.

b) S. AUBRY, P.Y. LE DAERON, G. ANDRE, Classical Ground States of a one-dimensional Model for Incommensurate Structures, Preprint Saclay, (1982).

32 S. AUBRY, G. ANDRE, Analyticity Breaking and Anderson localization in Incommensurate Lattices, Ann. of Israel Phys. Soc., 3 (1980) 133.

33 V.I. ARNOLD, a) Small divisors I. On the mapping of a circle into itself, Izv. Akad. Nauk, SSSR. Mat., 25 (1) 21-86,
b) Small divisors II : Proof of a A.N. KOLMOGOROV'S theorem on conservation of conditionnally periodic motion under small perturbations of the Hamiltonian function, Usp. Mat., Nauk, 18 (5) 13-40,
c) Small Divisors problem in classical and Celestial Mechanics, Usp. Mat., Nauk, 18 (6) 91-192.

34 J. AVRON, B. SIMON, Singular Continuous Spectrum for a Class of Almost periodic Jacobi Matrices, Bull. A.M.S., 6 (1982) 81-86.

35 J.E. BAYFIELD, P.M. KOCH, Multiphotonic Ionization of Highly Excited Hydrogen Atoms, Phys. Rev. Lett. 33 (1974) 258.

36 J.E. BAYFIELD, L.D. GARNER, P.M. KOCH, Observation of Resonances in the Microwave-Simulated Multiphotonic Excitation and Ionization of Highly Excited Hydrogen Atoms, Phys. Rev. Lett. 39 (1977) 76.

37 J. BELLISSARD, Almost Periodic Schrödinger Operators : an overview, Lecture Notes in Phys. 153 (1982) 356, Springer Verlag, Berlin, Heidelberg, New York.

38 J. BELLISSARD, Small divisors in Quantum Mechanics, in ref. [9].

39 J. BELLISSARD, R. LIMA, E. SCOPPOLA, Localization in -dimensional incommensurate structures, Comm. Math. Phys., 88 (1983) 465.

40 M. BERRY, in Ref. [8], [9], [24].

41 N.N. BOGOLJUBOV, N.M. KRYLOV, Sur quelques formules de développement en séries dans la mécanique non linéaire, Ukranin. Akad. Nauk. Kiev, 4 (1934) 56, see also Ann of Math. Studies II (1947) 106.

42 G. CASATI, I. GUARNERI, Chaos and Special Features of Quantum Systems under external perturbations, Phys. Rev. Lett. 50 (1983) 640.

43 G. CASATI, I. GUARNERI, Non Recurrent Behaviour in Quantum Dynamics, Comm. Math. Phys. 95 (1984) 121-127.

44 G. CASATI, B.V. CHIRIKOV, F.M. IZRAELEV, J. FORD, in Ref. [8].

45 B.V. CHIRIKOV, F.M. IZRAELEV, D.L. SHEPELYANSKI, in Soviet Scientific Review, Section C, Vol. 2 (1981).

46 W. CRAIG, Dense Pure Point Spectrum for the Almost Periodic Hill's equation, Comm. Math. Phys. 88 (1983) 113.

47 R.J. DAMBURG, V.V. KOLOSOV, A Hydrogen Atom in a Uniform Electric Field-III, J. Phys. B12 (1979) 2637-2643.

48 B. DORIZZI, B. GRAMMATICOS, Y. POMEAU, The periodically kicked Rotator : recurrence and/or energy growth, In Ref. [9].

49 D.F. ESCANDE, In Intrinsic Stochasticity in Plasmas, Ed. G. Laval and D.Gresillon Editions de Physique, Courtaboeuf (France) (1979).

50 D.F. ESCANDE, F. DOVEIL, Renormalization Method for the Onset of Stochasticity in a Hamiltonian System, Phys. Lett. 83A (1981) 307.

51 D.F. ESCANDE, F. DOVEIL, Renormalization Method for Computing the threshold of the large Scale Stochastic Instability in two degrees of freedom Hamiltonian Systems, J. Stat. Phys. 26 (1981) 257-284.

52 J.M. GREENE, A Method for determining a Stochastic transition, J. Math. Phys. 20 (1979) 1183-1201.

53 D.R. GREMPEL, S. FISHMAN, R.E. PRANGE, Localization in an incommensurate potential : an exactly Solvable model, Phys. Rev. Lett. 49 (1982) 833.

54 D.R. GREMPEL, S. FISHMAN, R.E. PRANGE, in Ref. [9].

55 E. HOPF, Statistik der geodätischen Linien in Mannig-faltigkesten negative Krümmung, Ber. Verh. Sächs. Akad. Wiss., Leipzig, 91 n° 3, (1939) 261-304.

56 J.S. HOWLAND, Stationnary Scattering theory for the time dependent Hamiltonians, Math. Ann. 207 (1974) 315-335.

57 F.M. IZRAELEV, D.L. SHEPELYANSKI, Quantum Resonnance for a Rotator in a Non Linear Periodic Field, Theor. Mat. Fiz., 43 (1980) 553-560, (English Transl.).

58 R. JENSEN, Stochastic Ionization of Surface State Electrons, Phys. Rev. Lett. 49 (1982) 1365.

59 R. JENSEN, Stochastic Ionization of Surface State Electrons : Classical Theory, Phys. Rev. A30 (1984) 386.

60 P.M. KOCH, Interactions of Intense Microwaves With Rydberg Atoms, J. Phys. (Paris) Colloq. C2 43 (1982) 187.

61 P.M. KOCH, D.R. MARIANI, Precise Measurements of the Static Electric Field Ionization Rate for Resolved Hydrogen Stark Substances, Phys. Rev. Lett. 46 (1981) 1275.

62 A.N. KOLMOGOROV, On the Conservation of Conditionnally periodic motion under small perturbations of the Hamiltonian function, Dokl. Akad. Nauk, SSSR, 98 (1954) 527-530.

63 J.G. LEOPOLD, I.C. PERCIVAL, Ionization of highly Excited Atoms by Electric Fields. III : Microwave ionization and Excitation, J. Phys. B12 (1979) 709-721.

64 J.N. MATHER, a) Existence of quasi-periodic Orbits for twist homeomorphisms of the Annulus, Topology 21 (1982) 457-467,
b) Non existence of invariant circles, Ergodic Theor. Dyn. Syst. To appear,
c) A Criterion for the Non Existence of Invariant Circles, Preprint Univ. Princeton, Math. Dpt., Aug. 9 (1982).

65 M. MORSE, Relation between the Critical Points of a Real Function of n independent variables, Transl. Amer. Math. Soc. 27 (1925) 345-396, Reprinted in Marston Morse's Selected Papers, R. Bott Ed. Springer Verlag (1981).

66 J. MOSER, On Invariant curves of a area preserving mapping of an Annulus, Nachr. Akad. Wiss. Göttingen. Math. Phys. : KI.NI.

67 I.C. PERCIVAL, Variational Principles for Invariant Tori and Cantori, In Symposium on Nonlinear Dynamics and Beam Beam Interactions, Amer. Inst. of Physics, Conf. Proc. n° 57, M. Month and J.C. Herrara Eds, (1980) 302-310, also J. Phys. A : 12 (1979) L57.

68 J. PÖSCHEL, Examples of Discrete Schrödinger operators with pure point Spectrum, Comm. Math. Phys. 88 (1983) 447.

69 K.F. ROTH, Rational Approximation to Algebraic Numbers, Mathematika 2 (1955) 1-20.

70 H. RÜSSMANN, On Optimal Estimates for the Solution of Linear Partial Differential Equations of first order with Constant Coefficients on the Torus, in Lecture Notes in Physics 38 (1975) 598-624, Springer Verlag, Berlin, Heidelberg, New York.

71 H. RÜSSMANN, Notes on sums Containing Small Divisors, Comm. Pure and Appl. Math. 29 (1976) 755-758.

72 H. RÜSSMANN, On Optimal Estimates for Solutions of linear difference equations on the Circle, Celestial Mechanics 14 (1976) 33-37.

73 H. RÜSSMANN, On the One-Dimensional Schrödinger Equation with a quasi periodic Potential, Ann., New York, Accad. Sci. 357 (1980) 90.

74 C.L. SIEGEL, Approximation algebraischer Zahlen, Math. Z. 10 (1921) 173-213.

75 C.L. SIEGEL, Iteration of Analytic Functions, Ann. Math. 43 (1942) 607-612.

76 Ya.G. SINAI, Dynamical Systems With Elastic Reflections, Russ. Math. Survey 25 (1970) 137-189.

77 K. SITNIKOV, Existence of oscillating motions for the three-body problem, Dokl. Akad. Nauk, USSR, 133 n° 2, (1960) 303-306.

78 J.B. TAYLOR, Unpublished (1968) see [52] .

79 A. THUE, Uber Annäherungswerte algebraischer Zahlen, J. Reine Angew. Math. 135 (1909) 284-305.

80 K. YAJIMA, Scattering theory for Schrödinger equations with Potential Periodic in time, J. Math. Soc. Japan 29 (1977) 729-743.

81 J. BELLISSARD, to appear in New trends in Mathematical Physics, Bielefeld 1982.

82 I thank G. CASATI and I. GUARNERI for this remark.

83 M. FEINGOLD, S. FISHMAN, D.R. GREMPEL, R.E. PRANGE, Statistics of Quasienergy Separations in Chaotic Systems, Preprint, Univ. of Maryland (1984).

Centre de Physique Théorique(Laboratoire propre, CNRS)*
Luminy - Case 907
13288 Marseille Cedex 9, France

*and Université de Provence

SOME RECENT APPLICATIONS OF STOCHASTIC PROCESSES IN QUANTUM MECHANICS

G. Jona-Lasinio
Dipartimento di Fisica Università di Roma "La Sapienza",
Roma, Italy.

1. There are essentially two ways in which the theory of stochastic processes has been used in connection with quantum mechanics. The most striking aspect is perhaps the (by now well established) mathematical equivalence of quantum mechanics to the theory of a special class of diffusion processes /1/. This is known as Stochastic Mechanics and in the last fifteen years has aroused a considerable interest also due to the possibility of constructing in this way an interpretation of Q.M. alternative to the standard one. A different way of connecting the theory of stochastic processes to Q.M. is via imaginary values of the time variable. This method reduces the Schrödinger equation to a heat equation which has a classical probabilistic interpretation. This approach, older than Stochastic Mechanics, has been mostly considered as a mathematical device to obtain certain properties of Schrödinger operators /2/. However there are cases where there is a direct relationship between the two methods.

2. In this Section we sketch some basic ideas of stochastic mechanics /3/. Consider the Schrödinger equation for a charged scalar particle in an electromagnetic field

$$i\hbar \partial_t \psi = \frac{1}{2m}\left(-i\hbar\nabla - \frac{e}{c}A\right)^2 \psi + V\psi \qquad (2.1)$$

A basic consequence of (2.1) is the continuity equation for the probability density $|\psi|^2$

$$\partial_t |\psi|^2 = -\nabla\left[\frac{\hbar}{m}\,\mathrm{Im}\left(\overline{\psi}\left(\nabla - \frac{ie}{\hbar c}A\right)\psi\right)\right] \qquad (2.2)$$

and this will be the starting point of our construction. Suppose we want to interpret $|\psi(x,t)|^2 = \rho(x,t)$ as the probability density at time t associated to a Markov process described by a transition function $P(x',x,t)$ in such a way that

$$\rho(x,t) = \int \rho_0(x')\, P(x',x,t)\, dx' \tag{2.3}$$

where $\rho_0(x)$ is some initial distribution. Then ρ must satisfy a Fokker-Planck equation

$$\partial_t \rho = \frac{\nu}{2} \Delta \rho - \nabla(b\rho) \tag{2.4}$$

where ν is the diffusion coefficient and b a velocity field (drift). Comparing (2.4) and (2.2) we see that such an identification is possible if we take $\nu = \hbar/m$ and b satisfying

$$\nabla\left[|\psi|^2\left(b - \frac{\hbar}{m}\frac{\mathrm{Im}\,(\bar{\psi}(\nabla - \frac{ie}{\hbar c}A)\psi)}{|\psi|^2}\right)\right] = \frac{\hbar}{2m}\Delta|\psi|^2 \tag{2.5}$$

The solution of (2.5) is not unique. However if we require that b is a gradient when A = 0 we obtain the unique answer

$$\begin{aligned} b &= \frac{\hbar}{m}\nabla \ln|\psi| + \frac{\hbar}{m}\frac{\mathrm{Im}\,(\bar{\psi}(\nabla - \frac{ie}{\hbar c}A)\psi)}{|\psi|^2} \\ &= \quad u \quad + \quad v \end{aligned} \tag{2.6}$$

Since u and v are given explicitely in terms of the wave function ψ, from the Schrödinger equation we obtain their equations of motion

$$\begin{aligned} \partial_t u &= -\frac{\hbar}{2m}\nabla(\nabla v) - \nabla(u\cdot v) \\ \partial_t v &= -\frac{1}{m}\nabla V + \frac{1}{2}\nabla(u^2 - v^2) + \frac{\hbar}{2m}\Delta u \end{aligned} \tag{2.7}$$

The stochastic process associated to b via (2.4) can be described also in terms of the stochastic differential equation

$$dx = b\,dt + \sqrt{\hbar/m}\;dw \tag{2.8}$$

where dw is the Wiener process.

At this point (2.7) and (2.8) (or (2.4)) constitute a self-contained scheme which can be used as an alternative to the usual quantum mechanical formalism.

For some time it was thought that it would be difficult to extend stochastic mechanics to particles with internal degrees of freedom e.g. to spin 1/2 particles obeying the Pauli equation(*)

$$i\partial_t \psi = \frac{1}{2}\left(-i\nabla - A\right)^2 + V\psi - \frac{1}{2}\,\underline{H}\cdot\underline{\sigma}\,\psi$$

$$\sigma_x = \begin{pmatrix} 0 & 1 \\ 1 & 0 \end{pmatrix}, \quad \sigma_y = \begin{pmatrix} 0 & -i \\ i & 0 \end{pmatrix}, \quad \sigma_z = \begin{pmatrix} 1 & 0 \\ 0 & -1 \end{pmatrix}, \quad \psi = \begin{pmatrix} \psi_1 \\ \psi_2 \end{pmatrix} \tag{2.9}$$

The reason for this belief was the lack of a classical analog for the spin. It turned out however that using the strategy previously indicated for the scalar case, the problem can be solved in a straightforward way simply by introducing discrete processes to describe the internal degrees of freedom /3/.

Consider for simplicity the case $\underline{H}$ = const.
Then we can write $\psi(x,t,\sigma) = \varphi(x,t)\chi(\sigma,t)$, $\sigma = \pm 1$ where φ satisfies the Schrödinger equation (2.1) and

$$i\,\frac{d\chi(\sigma)}{dt} = -\frac{1}{2}\left[H_z\sigma\chi(\sigma) + \left(H_x - i\sigma H_y\right)\chi(-\sigma)\right] \tag{2.10}$$

There is a continuity equation associated to (2.10)

$$\frac{d|\chi(\sigma)|^2}{dt} = \mathrm{Im}\left[\left(H_x + i\sigma H_y\right)\chi(\sigma)\overline{\chi}(-\sigma)\right] \tag{2.11}$$

(*) To simplify the subsequent discussion we set all the physical constants equal to 1.

The idea is the same as before: we try to interpret (2.11) as a discrete Fokker-Planck equation (Kolmogorov forward equation) for $\rho(\sigma) = |\chi(\sigma)|^2$

$$\frac{d\rho(\sigma)}{dt} = -p(\sigma,t)\rho(\sigma) + p(-\sigma,t)\rho(-\sigma) \tag{2.12}$$

where $p(\sigma,t)$ is a transition probability per unit time. The comparison of (2.12) and (2.11) gives the essentially unique choice for p

$$p(\sigma,t) = \frac{1}{2}\left\{[H_x^2 + H_y^2]^{1/2}\left|\frac{\chi(-\sigma)}{\chi(\sigma)}\right| + \mathrm{Im}(H_x - i\sigma H_y)\frac{\chi(-\sigma)}{\chi(\sigma)}\right\} \tag{2.13}$$

One can then easily find analogues of u and v

$$\begin{aligned} r &= H_z + \sigma\,\mathrm{Re}\left((H_x - i\sigma H_y)\chi(-\sigma)/\chi(\sigma)\right) \\ s &= -\sigma\,\mathrm{Im}\left((H_x - i\sigma H_y)\chi(-\sigma)/\chi(\sigma)\right) \end{aligned} \tag{2.14}$$

which obey the equations of motion

$$\begin{aligned} \frac{dr}{dt} &= -\sigma r s \\ \frac{ds}{dt} &= -\frac{1}{2}\sigma|H|^2 + \frac{1}{2}\sigma(r^2 - s^2) \end{aligned} \tag{2.15}$$

Notice the similarity in structure with (2.7) if one interprets ∇ as multiplication by σ. In terms of r and s the transition probability reads

$$p(\sigma,t) = \frac{1}{2}\left[(s^2 + (r - H_z)^2)^{1/2} - \sigma s\right] \tag{2.16}$$

Not all the solutions of (2.15) have the structure (2.14) so that an additional condition has to be imposed. For this point we refer reader to /3/.

A characteristic feature in the above construction is the appearence of the ratio $\chi(-\sigma)/\chi(\sigma)$. According to Cartan there is a geometric meaning associated to it. Introduce the so called isotropic vectors associated to spinors.

$$\begin{aligned} Z_1 &= \chi(\sigma)^2 - \chi(-\sigma)^2 \\ Z_2 &= i\left(\chi(\sigma)^2 + \chi(-\sigma)^2\right) \\ Z_3 &= -2\chi(\sigma)\chi(-\sigma) \end{aligned} \tag{2.17}$$

which satisfy the equation

$$Z_1^2 + Z_2^2 + Z_3^2 = 0 \tag{2.18}$$

This is a cone and the ratio $\chi(-\sigma)/\chi(\sigma)$ defines a generator of the cone. If we introduce now the complex variable Z = r + is equation (2.15) takes the very simple form

$$i\frac{dZ}{dt} = \frac{1}{2}\sigma|H|^2 - \frac{1}{2}\sigma Z^2 \tag{2.19}$$

which can be interpreted geometrically as an equation of motion for a generator of the isotropic cone.

The general case of an inhomogeneous magnetic field can be treated along similar lines. The picture which emerges is that a spin 1/2 particle can be described stochastically as a Brownian particle with two internal states moving in a velocity field which depends on the internal state. The latter changes at random times which in turn depend on the motion of the particle. The equations for the general case can be found in /3/.

3. A point of view /4/ very close to stochastic Mechanics turned out to be very useful in the study of the semiclassical limit of Q.M. /5/, i.e. the limit $\hbar/m \to 0$. This is based on the following remark. Suppose we want to study the spectrum of a Schrödinger operator without magnetic field. Then the ground state wave function ψ_o can be taken real and positive and the following unitary transformation is meaningful.

$$H \to \psi_o^{-1} H \psi_o = -\hbar L + E_o \tag{3.1}$$

where

$$\begin{aligned} L &= \frac{\hbar}{2} \Delta + b \nabla \\ b &= \hbar \nabla \ln \psi_o \end{aligned} \tag{3.2}$$

we have set m = 1.

The eigenvalues and eigenfunctions of H and L are therefore simply related by

$$\begin{aligned} E_K - E_o &= \hbar \lambda_K \\ \psi_K &= \psi_o \, \ell_K \end{aligned} \tag{3.3}$$

From (3.2) it follows that b coincides with the u defined by (2.6) for the ground state and satisfies

$$\hbar \, \mathrm{div}\, b + b^2 = 2(V - E_o) \tag{3.4}$$

which is the content of the second of (2.7) for this case.

The operator L is the generator of the Markov process associated by Stochastic Mechanics to the ground state. The interest of this transformation of the original problem lies in the fact that in many concrete cases eq. (3.4) can be solved with sufficient accuracy in the limit $\hbar \to 0$ while in the same limit powerful probabilistic methods can be applied to the analysis of the operator L.

For details on the application of this method we refer to /5/. Here we recall some of the striking results that were obtained in this way. It was proved for the first time that tunneling phenomena can be very sensitive to the detailed shape of the potential. As an example consider a potential like in Fig. 1

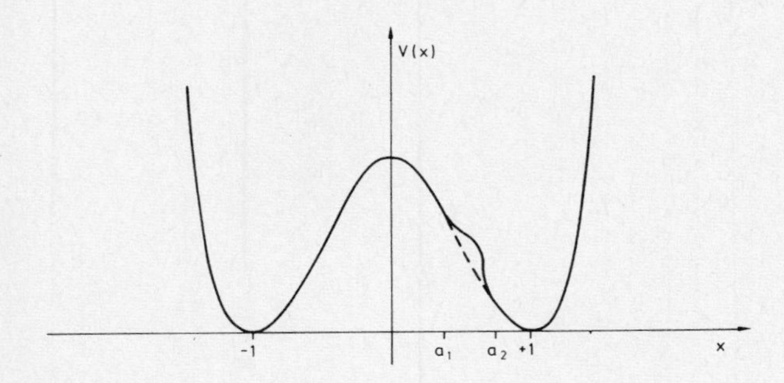

- Fig. 1 -

that is a symmetric double well locally perturbed on the interval $a_1 \leq x \leq a_2$. This is called by B.Simon the "flea on the elephant".

The flea is sufficient to destabilize tunneling to the point that the ground state wave function becomes exponentially localized in the first well

$$\psi_0(-1)/\psi_0(1) \approx \exp\left\{ \frac{1}{\hbar} \int_{-a_2}^{a_2} \sqrt{2V(x)}\, dx \right\} \tag{3.5}$$

while the splitting between the first excited state and the ground state becomes exponentially large if compared to the symmetric situation i.e.

$$\frac{(E_1 - E_0)_{\text{perturbed}}}{(E_1 - E_0)_{\text{symmetric}}} \approx \exp\left\{ \frac{2}{\hbar} \int_0^{a_2} \sqrt{2V(x)}\, dx \right\} \tag{3.6}$$

For other examples of instability of tunneling we refer the reader to /5/ and to the more recent papers /6/ /7/ /8/ where the same type of problems are treated by non probabilistic techniques.

This instability of tunneling is very relevant in understanding properties of disordered systems like Anderson localization /9/ and seems to provide a clue also to other phenomena like the existence of chiral molecules /10/.

4. In this Section we give another example where the point of view of Stochastic Mechanics has offered an interesting approach: the decay of resonant states /11/. This is a time-dependent problem and we need both eqs. (2.7). As we consider for simplicity a one-dimensional situation they simplify to

$$\begin{aligned} \partial_t u &= -\frac{\hbar}{2}\,\partial_x^2 v - \partial_x (uv) \\ \partial_t v &= -\partial_x V + \frac{1}{2}\,\partial_x \left(u^2 - v^2\right) + \frac{\hbar}{2}\,\partial_x^2 u \end{aligned} \tag{4.1}$$

The situation we want to discuss is the evolution of a state initially confined in a potential well whose minimum is higher than the rest of the world (see Fig. 2)

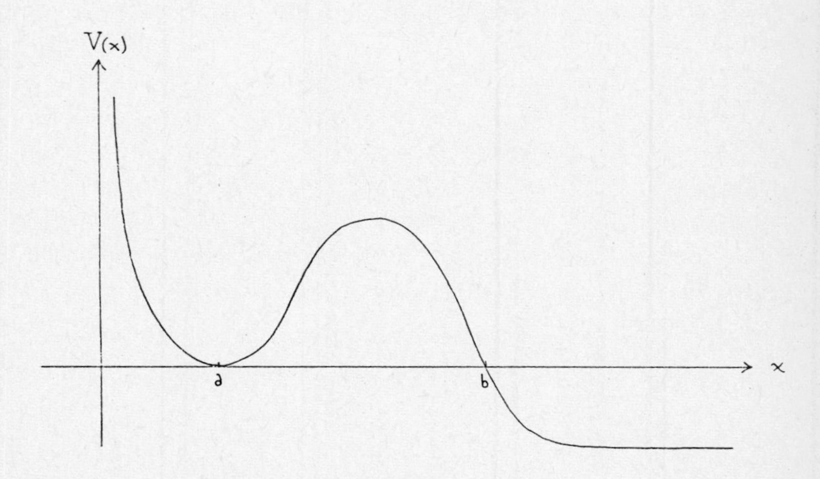

- Fig. 2 -

We proceed as follows. One first tries to construct initial data for (4.1) in such a way that the derivatives $\partial_t u$ and $\partial_t v$ at t = 0 are small, for example of the order $\hbar^\alpha$, $\alpha > 0$. We call initial data satisfying the above criterium quasi-stationary states. In our specific example the problem consists in constructing a quasi stationary state which corresponds to a wave function peaked around the point x = a. Our choice for the initial data is

$$u_0(x) = \begin{cases} \sqrt{2V(x)} & 0 \le x \le a \\ -\sqrt{2V(x)} & a \le x \le b - \epsilon(\hbar) \\ 0 & x > b + \epsilon(\hbar) \end{cases}$$

$$v_0(x) = \begin{cases} 0 & 0 \le x \le b - \epsilon(\hbar) \\ \sqrt{2|V(x)|} & x > b + \epsilon(\hbar) \end{cases} \tag{4.2}$$

In the small interval $(b - \epsilon(\hbar),\ b + \epsilon(\hbar))$ u_0 and v_0 are smoothed so as to be at least twice differentiable. It is then easy to see that for this choice the right hand sides of (4.1) are either zero or small with $\hbar$. The wave function ψ_0 corresponding to (4.2) is not normalizable but it is bounded at infinity and exponentially small in $\hbar$ if compared with its peak value at x = a. Our choice of v_0 corresponds to an outgoing wave. Of course one could work with normalizable data by putting the system in a box.

The next step consists in studying the stochastic differential equation (2.8) or the corresponding generator taking $b(x) = b_0(x) = u_0(x) + v_0(x)$. This approximation will be justified for times $t \le t_0$ such that the solutions of (4.1) remain close to u_0 and v_0. It is then natural to compute the probability P(t) that the process starting in a neighborhood escapes from the region (0,b) within time $t \le t_0$. Since in stochastic mechanics the process (2.8) describes the motion of the particle P(t) can be interpreted as the probability of decay of the quantum state within time t.

The calculation of P(t) can now be done using the theory of small random perturbations of dynamical systems /12/ and the details can be found in /11/. The result is that there exists a time interval $(\bar{t},\ t_0)$ over which the probability that the exit time τ of the process be greater than t is given by

$$P(\tau > t) \approx \exp\left\{-t\,c\,e^{-\frac{2}{\hbar}\int_a^b \sqrt{2V(x)}\,dx}\right\} \tag{4.3}$$

C may depend on $\hbar$ but the dependence is much slower than the exponential. (4.3) is the expected result but stochastic mechanics has the advantage over more traditional approaches of providing easily an indication on the interval of time during which the decay law is exponential. Quite generally it seems fair to say that stochastic mechanics is particularly suited to the qualitative analysis of the semiclassical limit.

5. In this last Section we briefly mention an extension of the imaginary time approach to equations of Pauli type /13/. The imaginary time Pauli equation reads

$$\partial_t \psi = -\frac{1}{2}\left(-i\nabla - A\right)^2 \psi - V\psi + \frac{1}{2}\,\underline{H}\cdot\underline{\sigma}\,\psi \tag{5.1}$$

We have seen in Section 2 that the stochastic description of the spin requires the introduction of jump processes. This suggests that in trying to generalize the well known Feynman-Kac formula to the Pauli equation one should consider expectations also with respect to a standard jump process (Poisson process) for a selected component of the spin.

This idea turns out to be correct and the solution of (5.1) can be written /13/

$$\begin{aligned}\psi(x,\sigma,t) &= e^{\lambda t}\, E\Big\{ e^{-\int_0^t V(x+w_s)\,ds \,-\, i\int_0^t A(x+w_s)\,dw_s} \\ &\cdot e^{\frac{1}{2}\int_0^t H_z(x+w_s)(-1)^{N_s}\sigma\,ds \,+\, \int_0^t \ln\left[\frac{1}{2\lambda}\left(H_x(x+w_s) - i\sigma(-1)^{N_s} H_y(x+w_s)\right)\right] dN_s} \\ &\cdot \psi_0\left(x+w_t,\, \sigma(-1)^{N_t}\right)\Big\}\end{aligned} \tag{5.2}$$

where N_t is the standard Poisson process of parameter λ . In the following we shall take $\lambda = 1$(*).

A representation like (5.2) is interesting because the algebra connected with the Pauli matrices has been replaced by the expectation with respect to the Poisson process and no chronological product is necessary. Generalizations can be written down easily for equations where the wave function has more then two components /13/.

There is a very simple inequality which follows from (5.2) by taking the absolute value

$$|\psi(x,\sigma,t)| \leq e^{t}\ E\Big\{ e^{-\int_0^t V ds + \frac{1}{2}\int_0^t H_z (-1)^{N_s}\sigma ds}$$

$$\cdot\, e^{\int_0^t \ln\left[\frac{1}{2}(H_x^2 + H_y^2)^{\frac{1}{2}}\right] dN_s}\ |\psi_0(x+w_t,\sigma(-1)^{N_t})|\Big\} \tag{5.3}$$

which implies the following inequality for the lowest eigenvalue of the Pauli Hamiltonian

$$E_0(A,\underline{H}) \geqslant E_0(0,\underline{H}') \tag{5.4}$$

where $\underline{H}' = \left(\sqrt{H_x^2 + H_y^2},\ 0,\ H_z\right)$

In words it compares the behaviour of a charged spinning particle in an arbitrary electromagnetic field with that of a neutral particle in a planar magnetic field. (5.3) and (5.4) generalize the so-called diamagnetic inequality for the Schrödinger equation /2/. As a final remark we observe that a representation of the form (5.2) may have a general mathematical interest in the study of the heat equation in the space of differential forms on a manifold.

(*) In concrete cases the arbitrariness of λ can be used to introduce a time scale natural for the problem.

References

/1/ E.Nelson: "Derivation of the Schrödinger Equation from Newtonian Mechanics", Phys. Rev. 150, 1079 (1966);
"Dynamical Theories of Brownian Motion", Princeton University Press 1967;
"Quantum Fluctuations" Princeton University Press 1984.

/2/ B.Simon: "Functional Integration and Quantum Physics", Acad. Press, New York 1979.

/3/ Here we follow the presentation in G.F.De Angelis, G.Jona-Lasinio: "A Stochastic Description of a Spin - 1/2 Particle in a Magnetic Field", J.Phys. A. 15, 2053 (1982).

/4/ S.Albeverio, R.Høegh-Krohn, L.Streit: "Energy Forms, Hamiltonian and Distorted Brownian Paths", J.Math. Phys. 18, 907 (1977).

/5/ G.Jona-Lasinio, F.Martinelli, E.Scoppola: "New Approach to the Semiclassical Limit of Quantum Mechanics", Comm. Math. Phys. 80, 223 (1981);
"The Semiclassical Limit of Quantum Mechanics: A Qualitative Theory via Stochastic Mechanics", Phys. Rep. 77, 313 (1981).

/6/ S.Graffi, V.Grecchi, G.Jona-Lasinio: "Tunnelling Instability via Perturbation Theory", J.Phys. A. 17, 2935 (1984).

/7/ B.Helffer, J.Sjöstrand: "Puits Multiples en Limite Semi-Classique" Preprint 1984.

/8/ B.Simon: "Semiclassical Analysis of Low Lying Eigenvalues, IV the Flea on the Elephant", Preprint 1984.

/9/ G.Jona-Lasinio, F.Martinelli, E.Scoppola: "Multiple Tunnellings in d-Dimensions: A Quantum Particle in a Hierarchical Potential", to appear in Annales de l'Institut Henri Poincaré.

/10/ P.Claverie, G.Jona-Lasinio: "Instability of Tunneling and the Concept of Molecular Structure in Quantum Mechanics: Some Remarks on the Enantiomer Problem", Preprint 1984.

/11/ G.Jona-Lasinio, F.Martinelli, E.Scoppola: "Decaying Quantum Mechanical States: An Informal Discussion within Stochastic Mechanics", Lett. Nuovo Cimento 34, 13 (1982).

/12/ A.D.Ventzel, M.I. Freidlin: "Dynamical Systems under the Action of Small Random Perturbations", Springer 1983.

/13/ G.F.De Angelis, G.Jona-Lasinio, M.Sirugue: "Probabilistic Solution of Pauli Type Equations", J.Phys. A 16, 2433 (1983).

The Quasi-Classical Limit of Scattering Amplitude
- Finite Range Potentials -

Kenji Yajima

Department of Pure and Applied Sciences
University of Tokyo
3-8-1 Komaba, Meguro-ku, Tokyo 153 Japan

§1. Introduction.

In these lectures the concern is with the asymptotic behavior as Planck's constant $h \to 0$ of the scattering operator S^h associated with the pair of time dependent Schrödinger equations

$$(1.1) \qquad i\hbar(\partial u/\partial t) = -(\hbar^2/2m)\Delta u + V(x)u \equiv H^h u ,$$

$$(1.2) \qquad i\hbar(\partial u/\partial t) = -(\hbar^2/2m)\Delta u \equiv H_0^h u .$$

We shall obtain the asymptotic expansion of the scattering amplitude $T^h(p,q)$ (in an average sense) and we shall prove the well-known conjecture that the limit as $h \to 0$ of the quantum mechanical total cross section is twice the one of classical mechanics. Here $m > 0$ is the mass parameter (hereafter $m=1$), $\hbar = h/2\pi$, $t \in \mathbb{R}^1$ (and $x \in \mathbb{R}^n$) is the time (and the space) variable, $\Delta = \partial^2/\partial x_1^2 + \ldots + \partial^2/\partial x_n^2$ and $V(x)$ is a real-valued function on $\mathbb{R}^n$. We assume here $V \in C_0^\infty(\mathbb{R}^n)$ for simplicity and refer to Yajima [13] for more general potentials.

Under this assumption the operators H^h and H_0^h are selfadjoint in the Hilbert space $L^2(\mathbb{R}^n)$ with the domain $D(H^h) = D(H_0^h) = H^2(\mathbb{R}^n)$ and the solution $u(t)$ to (1.1) (or (1.2)) with the initial condition $u(0) = u_0$ is given by $u(t) = \exp(-itH^h/\hbar)u_0$ (or $\exp(-itH_0^h/\hbar)u_0$). For studying the large time behavior of the solutions of (1.1) and the associated scattering theory one introduces the wave operators $W_\pm^h$:

$$(1.3) \qquad W_\pm^h = \operatorname*{s-lim}_{t\to\pm\infty} \exp(itH^h/\hbar)\exp(-itH_0^h/\hbar).$$

It is well-known (cf. Kuroda [5]) that these limits exist, $W_\pm^h$ are isometries and the images of W_+^h and W_-^h are identical with the continuous spectral subspace of H^h. The scattering operator S^h is then defined as $S^h = W_+^{h*}W_-^h$. S^h is a unitary operator on $L^2(\mathbb{R}^n)$ and the relation $S^h f_- = f_+$ is equivalent to

(1.4) $$\lim_{t\to\pm\infty} ||\exp(-itH^h/\hbar)f - \exp(-itH_0^h/\hbar)f_\pm|| = 0$$

with $f = W_+^h f_+ = W_-^h f_-$. It follows that S^h commutes with $\exp(-itH_0^h/\hbar)$ for all t, hence S^h and H_0^h may be simultaneously diagonalized. Since H_0^h is diagonalized via the Fourier transform

$$(\mathcal{F}^h f)(p) \equiv h^{-n/2} \int \exp(-ix\cdot p/\hbar)f(x)dx$$

as $(\mathcal{F}^h H_0^h \mathcal{F}^{h*} f)(p) = (p^2/2)f(p)$, we see that the distribution kernel $S^h(p,q)$ of the Fourier transform $\mathcal{F}^h S^h \mathcal{F}^{h*}$ of S^h is supported on $\{(p,q): p^2 = q^2\}$. It in fact is shown by means of the stationary theory of scattering (cf. Simon [8]) that $S^h(p,q)$ has the expression

$$S^h(p,q) = \delta(p-q) + \delta(p^2/2 - q^2/2)T^h(p,q)$$

with smooth $T^h(p,q)$. This function T^h is called the scattering amplitude.

Equation (1.1) (resp. (1.2)) is of course the quantization of the classical system with the Hamiltonian $H(x,p) = p^2/2 + V(x)$ (resp. $H_0(x,p) = p^2/2$) and it is the belief that any quantum quantity, in particular, the scattering amplitude $T^h(p,q)$ can asymptotically as $h \to 0$ be expressed in terms of the classical quantities.

We shall show in Sect. 5 that for fixed $q \in \mathbb{R}^n \setminus \{0\}$ $T^h(p,q)$ has an asymptotic expansion of the form $i(-i\hbar)^{-(n-1)/2} e^{i\phi(p)/\hbar} \sum_{j=0}^{\infty} \hbar^j \psi_j(p)$ in L^2-sense and compute the phase $\phi(p)$ and the first order term $\psi_0(p)$ in terms of the classical quantities. We outline here the argument which leads to this expansion, postponing the precise statements and proofs to the text. We denote the canonical maps induced by Hamilton equations

(1.5) $$\dot{x}(t) = \partial H/\partial p(x(t),p(t)), \quad \dot{p}(t) = -\partial H/\partial x(x(t),p(t))$$

by $\Gamma(t)$: $\Gamma(t)(x_0,p_0) = (x(t),p(t))$ being the solution of (1.5) with $x(0) = x_0$, $p(0) = p_0$. $\Gamma_0(t)(x_0,p_0) = (x_0+tp_0,p_0)$.

We first make the reduction of the problem. We take and fix a direction of the incoming momentum, say, x_1-direction. We write $x = (x_1,\underline{x}) \in \mathbb{R}^n$ with $x_1 \in \mathbb{R}^1$ and $\underline{x} \in \mathbb{R}^{n-1}$. Take a non-negative function $w(\underline{x}) \in C_0^\infty(\mathbb{R}^{n-1})$ such that $w(\underline{x}) = 1$ for $|\underline{x}| \leqq 1$ and $w(\underline{x}) = 0$ for $|\underline{x}| \geqq 2$ and $g(p_1) \in C_0^\infty((v_1, v_2))$ with $g(p_1) = 1$ for $v_3 < p_1 < v_4$, and set $w_R(\underline{x}) = w(\underline{x}/R)$ for $R > 1$, $f_R^h(x) = (\mathcal{F}^{h*}g)(x_1)w_R(\underline{x})$ and $\tilde{f}_R^h(x) = h^{-(n-1)/2}f_R^h(x)$. Since

$$(\mathcal{F}^h\tilde{f}_R^h)(q) \to g(q_1)\delta(\underline{q}) \quad \text{as} \quad R \to \infty,$$

and $T^h(p,q)$ is smooth, it follows that

$$(1.6) \qquad (\mathcal{F}^h(S^h-1)\tilde{f}_R^h)(p) = \int \delta(p^2/2-q^2/2)T^h(p,q)\mathcal{F}^h\tilde{f}_R^h(q)dq$$

$$\to |p|^{-1}T^h(p,|p|e_1), \quad R \to \infty,$$

e_1 being the unit vector in the x_1-direction. Thus the asymptotic expansion of $T^h(p,|p|e_1)$ as $h \to 0$ may be obtained from that of $\mathcal{F}^h(S^h-1)f_R^h$ which is uniform in R. We note that $f_R^h(x)$ is the function used by Enss-Simon [4] for defining their total cross section and it represents quasi-classically the ensemble of classical particles sitting on the Lagrangian plane $\{(x,\underline{p}): x_1=0,\ \underline{p}=0\}$ in the phase space $\mathbb{R}^{2n}$ with density $|g(p_1)|^2|w_R(\underline{x})|^2dp_1d\underline{x}$ (cf. Yajima [10]I, p.103).

To analyze $(S^h-1)f_R^h$ we argue as follows.

For large negative $t \leqq -T_1$, $\exp(-itH_0^h/\hbar)f_R^h$ is essentially supported in $x_1 < tv_1$. So the actions of $\exp(-isH^h/\hbar)$ and $\exp(-isH_0^h/\hbar)$ on it should coinside as far as $t+s < -T_1$ for all $R > 0$, that is

$$(1.7) \qquad \sup_{0<s\leqq-(T_1+t)} ||[\exp(-isH^h/\hbar)-\exp(-isH_0^h/\hbar)]\exp(-itH_0^h/\hbar)f_R^h|| \leqq C_Nh^N$$

with C_N independent of $R > 0$. Setting $s = -(T_1+t)$ and letting $t \to -\infty$ in (1.7), we are led to

$$(1.8) \qquad ||\exp(iT_1H^h/\hbar)W_-^hf_R^h - \exp(iT_1H_0^h/\hbar)f_R^h|| \leqq C_Nh^N.$$

We shall show this in Sec.2.

In Sec.3 we shall show that the large $|\underline{x}|$-part of f_R^h is irrelevant to the scattering: If $L >> 1$, then uniformly in $t > 0$ and $R \geqq 2L$,

$$(1.9) \quad ||[\exp(-itH^h/\hbar) - \exp(-itH_0^h/\hbar)]\exp(iT_1H_0^h/\hbar)(f_R^h-f_L^h)|| \leqq C_Nh^N.$$

Combing (1.8) and (1.9), we see

$$(1.10) \qquad \sup_{t \geqq 0} \| \exp(-i(t-T_1)H^h/\hbar)W_-^h f_R^h - \{\exp(-itH^h/\hbar)\exp(iT_1H_0^h/\hbar)f_L^h - \exp(-i(t-T_1)H_0^h/\hbar)(f_R^h - f_L^h)\}\| \leqq C_N h^N .$$

We then investigate the second term in the norm of (1.10) in Sec.4. For doing so, we need to assume the completeness of the classical scattering theory.

Assumption C. Any solution $(x(t)), p(t))$ of Hamilton equations (1.5) which is free: $x(t) = a + tve_1$, $p(t) = ve_1$ at $t \sim -\infty$ with $v \in (v_1, v_2)$ becomes free: $x(t) = a_+ + tp_+$, $p(t) = p_+$ again at $t \sim +\infty$.

Under this assumption we shall prove with large enough T_3 that

$$(1.11) \qquad \sup_{t>0} \|\{\exp(-itH^h/\hbar)-\exp(-itH_0^h/\hbar)\} \times \exp(-i(T_1+T_3)H^h/\hbar)\exp(iT_1H_0^h/\hbar)f_L^h\| \leqq C_N h^N.$$

Multiplying $\exp(i(t-T_3)H_0^h/\hbar)$ in the norm of (1.10), using relation (1.11) and then taking the limit $t \to \infty$, we obtain, uniformly in $R > 1$,

$$(1.12) \qquad \|\mathcal{F}^h(S^h-1)f_R^h - \mathcal{F}^h\{\exp(iT_3H_0^h/\hbar)\exp(-i(T_1+T_3)H^h/\hbar) \times \exp(iT_1H_0^h/\hbar)-1\}f_L^h\| \leqq C_N h^N.$$

This completely reduces the problem of finding the expansion of $\mathcal{F}^h(S^h-1)f_R^h$ which is uniform in R to that of R-independent $\exp(iT_3H_0^h/\hbar)\ \exp(-i(T_3+T_1)H^h/\hbar)\ \exp(iT_1H_0^h/\hbar)f_L^h$ which is an object of finite time. We shall prove in Sect.5 that the latter has an asymptotic expansion to any order in $\hbar$ at any outgoing momentum where the classical differential cross section has no singularity. Obviously this proves the desired expansion of $T^h(p,q)$. In sections 4 and 5, Fujiwara's approximate fundamental solution which will be used to replace $\exp(-i(T_1+T_3)H^h/\hbar)$ by an oscillatory integral operator (with a very large number of variables) and the method of stationary phase play the central roles.

In the final Sect.6, we shall discuss a well-known conjecture [4] on the limit as $h \to 0$ of the total cross section

$$(1.13) \qquad \lim_{h \to 0} h^{(n-1)}\sigma_{tot}^h(\lambda) = 2\sigma_{tot}(\text{classical})$$

and give a sufficient condition for (1.13) to hold that the incoming and outgoing waves do not interfare each other. We also show (1.13) will in general break down if this condition is not satisfied.

The formal expansion formula for $T^h(p,q)$ may be found in many physics works (cf. Miller [7] and the references there in). However the only rigorous work which we know of is of Vainberg [9] who proved it in compact uniform topology but only for large $|p| = |q|$ and for potentials of compact support. Our results extend that of [9] by removing the energy restriction and our method gives access to treat more general potentials which may not have compact supports (cf. [12]). The asymptotic behavior of S^h applied to a class of oscillating wave functions were studied in a series of papers by Yajima [10], [11]. Various inspections on (1.13) is given by Enss and Simon [4].

Some of the notations. $L^p(\mathbb{R}^n)$ is the Banach space of p-summable functions. The norm of $L^p(\mathbb{R}^n)$ will be written $\|\ \|_p$ regardless of the dimension n. The norm of the Sobolev space of order s, $H^s(\mathbb{R}^n)$, will be denoted as $\|\cdot\|_{H^s}$. The Fourier transform with $h = 2\pi$ will be denoted $\hat{\omega}(p) = (\mathcal{F}^h\omega)(p)|_{h=2\pi}$.

§2. Large negative time behavior.

In this section we show the inequality (1.7) or (1.8) stated in Sect.1. We begin with the following lemma. $0 < v_1 < v_2 < \infty$ are fix constants and

$$K_0^h = -(\hbar^2/2)d^2/dx_1^2,\ \underline{K}_0^h = -(\hbar^2/2)(\partial^2/\partial x_2^2 + \dots + \partial^2/\partial x_n^2).$$

Lemma 2.1. Let $g \in C_0^\infty((v_1, v_2))$ and $a > 0$. Then for any positive integer N there exists a constant C_N such that for any $|t| \geq (2a+v_1)/v_1 = T_1$, $|x_1| < a$ and $0 < h \leq 1$

$$(2.1)\qquad |\exp(-itK_0^h/\hbar)\mathcal{F}^{h*}g(x_1)| \leq C_N h^N(1+|t|)^{-N}$$

Proof In the RHS of

$$\exp(-itK_0^h/\hbar)\mathcal{F}^{h*}g(x_1) = h^{-1/2}\int_{-\infty}^{\infty}\exp(i(x_1p-tp^2/2)/\hbar)g(p)dp$$

the phase function satisfies

$$|(d/dp)(x_1p-tp^2/2)| = |x_1-tp| \geq |t|v_1-a \geq \tfrac{1}{2}(|t|+1)v_1$$

for $|t| \geq (2a+v_1)/v_1$, $|x_1| < a$ and $p \in (v_1, v_2)$. Hence $(N+1)$ integrations by parts yield

$$|(e^{-itK_0^h/\hbar}\mathcal{F}^{h*}g(x_1)| = |h^{-1/2}\int_{-\infty}^{\infty}\{\frac{(-i\hbar)}{(x_1-tp)}\frac{\partial}{\partial p}\}^N e^{i(x_1p-tp^2/2)/\hbar}g(p)dp|$$

$$= (2\pi)^{-1/2}\hbar^{N+\frac{1}{2}} \int_{\infty}^{\infty} |\{\frac{1}{(x_1-tp)}\frac{\partial}{\partial p}\}^{*N} g(p)|dp$$

$$\leq C_N h^N (1+|t|)^{-N} .$$

Lemma 2.2. For any $t \in \mathbb{R}$

$$||\exp(-it\underline{K}_0^h/\hbar)w_R||_{\infty} \leq (2\pi)^{-(n-1)/2}||\hat{w}||_1. \tag{2.2}$$

Proof. After a simple manipulation we have

$$\exp(it\underline{K}_0^h/\hbar)w_R(\underline{x}) = (2\pi)^{-(n-1)/2}\int_{\mathbb{R}^{n-1}} e^{i(\underline{x}\cdot\underline{p}-t\underline{p}^2/2)/\hbar} (R/\hbar)^{n-1}\hat{w}(R\underline{p}/\hbar)d\underline{p}.$$

The estimate (2.2) immediately follows from this.

Let us suppose that a is large and

$$\text{supp } V \subset \{x \in \mathbb{R}^n: |x| < a\}. \tag{2.3}$$

We set $f_R^h(x) = (\overset{*}{\mathcal{F}}{}^h g)(x_1)w_R(\underline{x})$. It follows from (2.1) and (2.2) that for $|x| \leq a$ and $|t| \geq T_1$,

$$|e^{-itH_0^h/\hbar} f_R^h(x)| = |e^{-itK_0^h/\hbar}\mathcal{F}^{h*}g(x_1)||e^{-it\underline{K}_0^h/\hbar}w_R(\underline{x})| \tag{2.4}$$

$$\leq C_N h^N(1+|t|)^{-N}.$$

Proposition 2.3. For any N there exists a constant C_N such that for any $R > 0$, $0 < h \leq 1$, $t \leq -T_1$ and $0 \leq s \leq -T_1-t$,

$$||(e^{-isH^h/\hbar}-e^{-isH_0^h/\hbar})e^{-itH_0^h/\hbar} f_R^h|| \leq C_N h^N , \tag{1.7}$$

in particular,

$$||\exp(iT_1H^h/\hbar)W_-^h f_R^h - \exp(iT_1H_0^h/\hbar)f_R^h|| \leq C_N h^N . \tag{1.8}$$

Proof. By Duhamel's identity, Minkowski's inequality and the unitarity of $\exp(-itH^h/\hbar)$, we have

$$\sup\{\text{LHS of (1.7)}: 0 \leq s \leq -T_1-t,\ t \leq -T_1\} \tag{2.5}$$

$$\leq \sup_{t\leq -T_1} \hbar^{-1}\int_0^{-(T_1+t)} ||Ve^{-i(t+s)H_0^h/\hbar} f_R^h|| \, ds$$

$$= \hbar^{-1}\int_{-\infty}^{-T_1} ||V\exp(-itH_0^h/\hbar)f_R^h|| \, dt.$$

Apply (2.3) and (2.4) to estimate the last member in (2.5) by

$$h^{-1}\int_{-\infty}^{-T_1} ||V||_2 C_N h^N(1+|t|)^{-N} \, dt$$

which leads to the conclusion of the proposition.

§3. Large $|\underline{x}|$-parts are irrelevant to scattering.

We fix the negative time $-T_1$ of Sect.2. We shall show here if L is sufficiently large, the part of f_R^h which lives outside $|\underline{x}| \geq 2L$ is irrelevant to the scattering. We write $\bar{w}_\varepsilon(\underline{p}) = 1-w_\varepsilon(\underline{p})$.

Lemma 3.1. Let $1 \leq 2L$ and $\varepsilon > 0$. Then for any $N > 0$, there exists a constant $C_{N,\varepsilon,L}$ independent of $R \geq 2L$ and $0 < h \leq 1$ such that

$$(3.1) \qquad ||\bar{w}_\varepsilon(\underline{p})\{\mathcal{F}^h\bar{w}_L w_R\}(\underline{p})||_2 \leq C_{N,\varepsilon,L}\, h^N .$$

Proof. Integrations by parts yield

$$(3.2) \qquad |h^{-(n-1)/2}\int\bar{w}_\varepsilon(\underline{p})\exp(-i\underline{x}\cdot\underline{p}/\hbar)\bar{w}_L(\underline{x})w_R(\underline{x})d\underline{x}|$$

$$= |h^{-(n-1)/2}\hbar^{2N}\int e^{-i\underline{x}\cdot\underline{p}/\hbar}\bar{w}_\varepsilon(\underline{p})|\underline{p}|^{-2N}(-\Delta_{\underline{x}})^N\{\bar{w}_L(\underline{x})w_R(\underline{x})\}d\underline{x}|$$

$$\leq \hbar^{2N}h^{-(n-1)/2}\bar{w}_\varepsilon(\underline{p})|\underline{p}|^{-2N}||(-\Delta_{\underline{x}})^N\bar{w}_L(\underline{x})w_R(\underline{x})||_1.$$

We note that in the RHS of the Leibnitz' rule

$$\partial^\alpha\bar{w}_L(\underline{x})w_R(\underline{x}) = \sum_{0\leq\beta\leq\alpha}\binom{\alpha}{\beta}\partial^{\alpha-\beta}\bar{w}_L(\underline{x})\partial^\beta w_R(\underline{x})$$

all the terms except $\bar{w}_L(\underline{x})\partial^\alpha w_R(\underline{x})$ and $w_R(\underline{x})\partial^\alpha\bar{w}_L(\underline{x})$ vanish since $\partial w_R/\partial x_j$ and $\partial\bar{w}_L/\partial x_j$ have disjoint supports. Hence

$$(3.3) \qquad ||(-\Delta_{\underline{x}})^N\bar{w}_L(\underline{x})w_R(\underline{x})||_1 \leq (L^{-2N+n-1}+R^{-2N+n-1})||\Delta^N\underline{w}||_1.$$

Combining (3.2) and (3.3), we obtain (3.1).

Proposition 3.2. Let $L \geq 3(T_1+a)$ and $R \geq 2L$. Then for any $N \geq 1$, there exists a constant $C_{N,L}$ independent of $R \geq 2L$ and $0 < h \leq 1$ such that

$$(3.4) \qquad \sup_{t\geq 0}||\{e^{-itH^h/\hbar}-e^{-itH_0^h/\hbar}\}e^{iT_1H_0^h/\hbar}(f_R^h-f_L^h)|| \leq C_{N,L}h^N.$$

Proof. By Duhamel's identity, Minkowski's inequality and the unitarity of $\exp(-itH^h/\hbar)$, the LHS of (3.4) is not larger than

$$(3.5) \qquad \hbar^{-1}\int_{-T_1}^\infty||V\exp(-itH^h/\hbar)(f_R^h-f_L^h)||dt$$

$$= \hbar^{-1}\int_{-T_1}^\infty||V(e^{-itK_0^h/\hbar}\mathcal{F}^{h*}g)(x_1)(e^{-it\underline{K}_0^h/\hbar}\bar{w}_L w_R)(\underline{x})||dt$$

$$= \hbar^{-1}\{\int_{-T_1}^{T_1} + \int_{T_1}^{\infty} \| \;\text{"}\; \| dt\} = I_1 + I_2.$$

By Lemma 2.1 and 2.2, (3.6) $\quad I_2 \leqq \hbar^{-1}\int_{T_1}^{\infty} C_N h^N (1+t)^{-N} dt =$

$C_N h^{N-1}(1+T_1)^{-N+1}$.

To estimate I_1, we observe

$$(3.7) \qquad \| V(e^{-itK_0^h/\hbar}\mathcal{F}^{h*}g)(\underline{x}_1)(e^{-it\underline{K}_0^h/\hbar}\bar{w}_L w_R)(\underline{x})\|^2$$

$$\leqq \|V\|_\infty^2 \; \|g\|_2^2 \; \|e^{-it\underline{K}_0^h/\hbar}\bar{w}_L w_R(\underline{x})\|^2_{L^2(|\underline{x}|\leqq a)}$$

$$\leqq 4\|V\|_\infty^2 \; \|g\|_2^2 (\|\bar{w}(\underline{p})\mathcal{F}^h \bar{w}_L w_R\|_2^2 +$$

$$+ \|e^{-it\underline{K}_0^h/h}\mathcal{F}^{h*} w(\underline{p})\mathcal{F}^h \bar{w}_L w_R\|^2_{L^2(|\underline{x}|\leqq a)}).$$

Here we used the unitarity of $\exp(-itH_0^h/\hbar)$ and $\mathcal{F}^h$ and the decomposition $(w(\underline{p})+\bar{w}(\underline{p}))\mathcal{F}^h \bar{w}_L w_R(\underline{p}) = \mathcal{F}^h \bar{w}_L w_R(\underline{p})$. For $|\underline{x}| \leqq a$ and $|t| < T_1$, the phase function in

$$e^{-it\underline{K}_0^h/\hbar}\mathcal{F}^{h*} w(\underline{p})\mathcal{F}^h \bar{w}_L w_R(\underline{x})$$

$$= h^{-(n-1)}\int e^{i(\underline{x}\cdot\underline{p}-t\underline{p}^2/2-\underline{y}\cdot\underline{p})/\hbar} \; w(\underline{p})\bar{w}_L w_R(\underline{y})d\underline{y}d\underline{p}$$

satisfies for $p \in \operatorname{supp} w$ and $y \in \operatorname{supp} \bar{w}_L w_R$

$$|\partial/\partial\underline{p}(\underline{x}\cdot\underline{p}-t\underline{p}^2/2-\underline{y}\cdot\underline{p})| = |\underline{x}-t\underline{p}-\underline{y}|$$
$$\geqq \tfrac{1}{3}|\underline{y}|+(\tfrac{2}{3}|\underline{y}|-|\underline{x}|-|t\underline{p}|) \geqq \tfrac{1}{3}|\underline{y}|+(T_1+a).$$

Hence by the integration by parts, we see that for $|\underline{x}| \leqq a$, $|t| \leqq T_1$

$$(3.8) \qquad |e^{-it\underline{K}_0^h/\hbar}\mathcal{F}^{h*} w(\underline{p})\mathcal{F}^h \bar{w}_L w_R(\underline{x})|$$

$$= |h^{-(n-1)}\hbar^N \int d\underline{p}d\underline{y} e^{i(\underline{x}\cdot\underline{p}-t\underline{p}^2/2-\underline{y}\cdot\underline{p})/\hbar}$$
$$\{\frac{-i}{(\underline{x}-t\underline{p}-\underline{y})}\cdot\frac{\partial}{\partial\underline{p}}\}^{*N} w(\underline{p})\overline{w_L}(\underline{y})w_R(\underline{y})|$$

$$\leqq h^{N-(n-1)}C'_{N,T_1,a}\int d\underline{p}d\underline{y}\{(1+|\underline{y}|)^{-N} \sum_{|\alpha|\leqq N}|\partial_{\underline{p}}^{\alpha} w(\underline{p})|\}.$$

Combining (3.7) and (3.8) with Lemma 3.1, we see

$$(3.9) \qquad I_1 \leq \hbar^{-1}\int_{-T_1}^{T_1}\{C_{N,1,L}h^N + C'_{N,T_1,a}h^{N-(n-1)}\}dt.$$

(3.5), (3.6) and (3.9) clearly imply the proposition.

Hereafter a and L are fixed as in (2.3) and in Proposition 3.2.

§4. Approximate fundamental solution and far future behavior

We assume here the condition C, the completeness of classical scattering theory. A simple compactness argument implies the following lemma of which the proof is omitted.

Lemma 4.1. Let $K \subset \mathbb{R}^{2n}$ be a compact subset such that $(x,p) \in K$ implies $x \notin \text{supp } V$ and $|p| > \delta > 0$ with $\delta > 0$ independent of (x,p). Then there exists $T_2 > 0$ such that $(x(t),p(t)) = \Gamma(t)(x_0,p_0)$ with $(x_0,p_0) \in K$ satisfies $x(t) \notin \text{supp } V$ for all $t \geq T_2$.

We denote $((0,\underline{x}), (v,\underline{0}))$ by $(\underline{x},v)^*$ for $(\underline{x},v) \in \mathbb{R}^{n-1} \times \mathbb{R}^1$ and $G^* = \{(\underline{x},v)^*: (\underline{x},v) \in G\} \subset \mathbb{R}^n \times \mathbb{R}^n$ for $G \subset \mathbb{R}^{n-1} \times \mathbb{R}^1$. $D = \{(\underline{x},v) \in \mathbb{R}^{n-1} \times \mathbb{R}: |\underline{x}| \leq 2L,\ v_1 \leq v \leq v_2\}$ and $K_{-1} = \Gamma_0(-T_1)D^*$. $\Omega(t) = \Gamma(t+T_1)\Gamma_0(-T_1)$.

By virtue of Lemma 4.1 we see that there exists T_2 such that all solutions $(x(t,\underline{a},v),\ p(t,\underline{a},v)) = \Gamma(t+T_1)\Gamma_0(-T_1)(\underline{a},v)^*$, $(\underline{a},v) \in D$, of (1.5) with the condition

$$(4.2) \qquad x(-T_1,\underline{a},v) = (-T_1 v,\underline{a}),\quad p(-T_1,\underline{a},v) = (v,0),$$

are all outside of supp V for $t > T_2$, hence

$$(4.3) \qquad x(t,\underline{a},v) = (t-T_2)p(T_2,\underline{a},v) + x(T_2,\underline{a},v),\quad p(t,\underline{a},v) = p(T_2,\underline{a},v).$$

Since

$$(4.4) \qquad |\,p(t,\underline{a},v)|^2 = |p(T_2,\underline{a},v)|^2 = v^2 > v_1^2\ ,$$

(4.3) implies the existence of T_3 such that for all $t \geq T_3$,

$$(4.5) \qquad \cos(x(t,\underline{a},v),\ p(t,\underline{a},v)) \geq 3/4,\quad |x(t,\underline{a},v)| \geq 10L.$$

We take $\chi \in C_0^\infty(\mathbb{R}^{2n})$ such that $\chi(x,p) = 1$ on a neighbourhood of $\Omega(T_3)D^*$ and $\chi(x,p) = 0$ if either one of $|p| < v_1/2$, $|x| < 5L$ and $\cos(x,p) \leq 1/2$ is satisfied. $\tilde{\chi}^h(y,D)$ is the (conjugate) pseudo-differential operator

$$(4.6) \qquad (\mathcal{F}^h\tilde{\chi}^h(y,D)f)(p) = h^{-n/2}\int e^{-iy\cdot p/\hbar}\chi(y,p)f(y)dy.$$

Proposition 4.2 Let T_3 and $\tilde{\chi}^h(y,D)$ be as above. Then for any N, there exists a constant C_N such that for all $0 < h \leq 1$,

$$(4.7)\qquad ||(1-\tilde{\chi}^h(y,D))\exp(-i(T_3+T_1)H^h/\hbar)\exp(iT_1H_0^h/\hbar)f_L^h|| \leq C_N h^N.$$

For the proof we replace $\exp(-i(T_3+T_1)H^h/\hbar)$ in (4.7) by an oscillatory integral operator by means of Fujiwara's approximate fundamental solution $E_N^h(t)$. Since we systematically use this $E_N^h(t)$, we state here some of its properties which will be strictly necessary in what follows, referring to Fujiwara [3] for the details.

If T_0 is sufficiently small, then for any $0 < T < T_0$ and any pair (x,y) of points in $\mathbb{R}^n$, there exists a unique solution $(x(t), p(t))$ of (4.1) with $x(0) = y$, $x(T) = x$. $S(T,x,y)$ is the action integral of this path:

$$S(T,x,y) = \int_0^T (p(t)^2/2 - V(x(t)))dt.$$

$$(4.8)\qquad \partial S/\partial x(T,x,y) = p(T), \quad \partial S/\partial y(T,x,y) = -p(0)$$

We set

$$e_N^h(T,x,y) = (1/2\pi iT)^{n/2} \sum_{j=1}^{N} \hbar^{j-1} a_j(T,x,y),$$

$$a_1(T,x,y) = \exp(-\tfrac{1}{2}\int_0^T \{(\Delta_x S)(t,x(t),y) - \tfrac{n}{2t}\}dt),$$

$$a_j(T,x,y) = -\tfrac{1}{2}a_1(T,x,y)\int_0^T \frac{\Delta_x a_{j-1}(t,x(t),y)}{a_1(t,x(t),y)}\,dt,\ j=2,\ \ldots,$$

and define $E_N^h(T)$ as an oscillatory integral operator:

$$E_N^h(T)f(x) = \hbar^{-n/2}\int e_N^h(T,x,y)e^{iS(T,x,y)/\hbar}f(y)dy.$$

This $E_N^h(T)$ satisfies for $0 < h \leq 1$ and $0 < |h| < T_0$,

$$(E.1)\qquad ||E_N^h(T)|| \leq C, \quad \operatorname*{s-lim}_{T\to 0} E_N^h(T) = 1$$

$$(E.2)\qquad ||E_N^h(T) - \exp(-iTH^h/\hbar)|| \leq C_N h^N T^{N-1}.$$

(E.3) Any power $E_N^h(T)^\ell$ is an Oscillatory Integral Operator of Asada-Fujiwara typ (AF-Os.IO, cf. [2]).

$$(4.9)\qquad E_N^h(T)^\ell f(x) = \hbar^{-n\ell/2}\int e_N^h(T,x,x_{\ell-1})e_N^h(T,x_{\ell-1},x_{\ell-2})\cdots$$

$$e_N^h(T,x_1,y) \times \exp\{i(S(T,x,x_{\ell-1})+S(T,x_{\ell-1},x_{\ell-2})+\cdots$$

$$+S(T,x_1,y))/\hbar\}f(y)dydx_1\cdots dx_{\ell-1}$$

$$= \hbar^{-n\ell/2}\int A_N^h(\ell T,x,\theta,y)e^{i\Phi(\ell T,x,\theta,y)/\hbar}f(y)dyd\theta,$$

$$\theta = (x_1, \dots, x_{\ell-1})$$

with

(i) $A_N^h(\ell T, x, \theta, y) \in \mathcal{B}(\mathbb{R}^{(\ell+1)n})$, uniformly for $0 < h \leq 1$,

(ii) the $\ell n \times \ell n$-matrix

$$D(\phi)(x,\theta,y) = \begin{bmatrix} \partial^2\Phi/\partial\theta\partial y(\ell T,x,\theta,y) & \partial^2\Phi/\partial\theta\partial y(\ell T,x,\theta,y) \\ \partial^2\Phi/\partial x\partial\theta(\ell T,x,\theta,y) & \partial^2\Phi/\partial\theta\partial\theta(\ell T,x,\theta,y) \end{bmatrix}$$

satisfies

(4.10) $\quad |\det D(\phi)(x,\theta,y)| \geq \delta > 0, \ (x,\theta,y) \in \mathbb{R}^{(\ell+1)n};$

(4.11) $\quad D(\phi)(x,\theta,y) \in \mathcal{B}(\mathbb{R}^{(\ell+1)n}).$

Proof of Proposition 4.2. We write $T_4 = T_3 + T_1$. Take a positive integer ℓ large enough so that $T = T_4/\ell < T_0$ and we may construct $E_N^h(T)$. By (E.1) and (E.2) we have

(4.12) $\quad ||\exp(-iT_4H^h/\hbar) - E_N^h(T)^\ell|| \leq C_N h^N.$

Using the expression (4.9) for $E_N^h(T)^\ell$, we write

(4.13) $$\mathcal{F}^h(1-\tilde{\chi}^h(y,D))E_N^h(T)^\ell \exp(iT_1H_0^h/\hbar)f_L^h(p)$$
$$= (2\pi)^{-(3n-1)/2}\hbar^{-n(\ell+3)/2+1/2}$$
$$\int \exp\{i(-x\cdot p+\Phi(x,\theta,z)+z\cdot q+iT_1q^2/2-\underline{q}\cdot\underline{y})/\hbar\}$$
$$(1-\chi(x,p))A_N^h(T_4,x,\theta,z)g(q_1)w_L(\underline{y})d\underline{y}dqdzd\theta dx$$

as an AF-OsIO. We write the phase function as $\mu(p,x,\theta,z,q,\underline{y})$. Suppose that $|\nabla_{(x,\theta,z,q,\underline{y})}\mu(p,x,\theta,z,q,\underline{y})| < \gamma_0$ for sufficiently small $\gamma_0 > 0$, $q_1 \in \operatorname{supp} g$ and $\underline{y} \in \operatorname{supp} w_L$. This will imply

$(4.14)_{-1}$ $\quad |\partial_{\underline{y}}\mu| = |\underline{q}| < \gamma_0, \quad |\partial_{q_1}\mu| = |z_1+T_1q_1| < \gamma_0$

$\quad |\partial_{\underline{q}}\mu| = |\underline{z}+T_1\underline{q}-\underline{y}| < \gamma_0$;

$(4.14)_0$ $\quad |\partial_z\mu| = |\partial S/\partial z(T,x_1,z)+q| < \gamma_0;$

$(4.14)_j$ $\quad |\partial_{x_j}\mu| = |\partial S/\partial x_j(T,x_{j+1},x_j)+\partial S/\partial x_j(T,x_j,x_{j-1})| < \gamma_0;$

$(4.14)_\ell$ $\quad |\partial_x\mu| = |\partial S/\partial x(T,x,x_{\ell-1})-p| < \gamma_0.$

Relations $(4.14)_{-1}$ imply that (q,z) is in γ_0-neibourhood K_0 of $K'_{-1} = \Gamma_0(-T_1)D^{*'}$, $D^{*'}$ being γ_0-neibourhood of D^*; $(4.14)_0$ implies $(x_1, \partial S/\partial x_1(T,x_1,z)) \in \Gamma(T)K'_0 = K_1$, K'_0 being γ_0-neibourhood of K_0;

$(4.14)_1$ implies $(x_2, \partial S/\partial x_2(T,x_2,x_1)) \in \Gamma(T)K_1'$, K_1' being γ_0-neibourhood of K_1 etc. Thus if γ_0 is sufficiently samll, $(4.14)_{-1} \sim (4.14)_\ell$ imply (x,p) is in a small neibourhood of $\Omega(T_3)D^*$ on which $\chi(x,p) = 1$. Thus we see that for some $\gamma > 0$,

$$|\nabla_{(x,\theta,z,q,\underline{y})}\mu(p,x,\theta,z,q,\underline{y})| > \gamma > 0$$

for all $(x,p) \in \mathrm{supp}(1-\chi)$, $q_1 \in \mathrm{supp}\, g$ and $|\underline{y}| < 2L$. Since the second derivatives of $\mu(p,x,\theta,z,q,\underline{y})$ are in $\mathcal{B}(\mathbb{R}^{(\ell+4)n-1})$, we have by integration by parts in $(x,\theta,z,q,\underline{y})$-variables and the L^2-boundedness theorem for AF-OsIO

$$||\mathcal{F}^h(1-\tilde{\chi}(y,D))E_N^h(T)^\ell \exp(iT_1H_0^h/\hbar)f_L^h||_2 \leq C_N\hbar^N||g\cdot w_L||_{H^N}.$$

Once Proposition 4.2 is proved, the standard technique of scattering theory shows that $\exp(-itH^h/\hbar)\exp(iT_1H_0^h/\hbar)f_L^h$ behaves essentially freely for $t \geqq T_3$.

<u>Proposition</u> 4.3. Let $T_4 = T_1 + T_3$ be as above. Then for any N, there exists C_N such that for $0 < h \leq 1$,

$$(4.15) \qquad \sup_{t\geqq 0} ||(e^{-itH^h/\hbar} - e^{-itH_0^h/\hbar})e^{-iT_4H^h/\hbar}e^{iT_1H_0^h/\hbar}f_L^h|| \leq C_N h^N$$

<u>Proof</u>. By virtue of (4.7), it suffices to show

$$(4.16) \qquad \sup_{t\geqq 0} ||(e^{-itH^h/\hbar} - e^{-itH_0^h/\hbar})\tilde{\chi}^h(y,D)d_L^h|| \leq C_N h^N$$

for $d_L^h = \exp(-iT_4H^h/\hbar)\exp(iT_1H_0^h/\hbar)f_L^h$. By Duhamel's identity, Minkowski's inequality and the unitarity of $\exp(-itH^h/\hbar)$, we have

$$(4.17) \qquad (\mathrm{LHS})\text{ of } (4.16) \leq \hbar^{-1}\int_0^\infty ||V\exp(-itH_0^h/\hbar)\tilde{\chi}^h(y,D)d_L^h||dt.$$

For $x \in \mathrm{supp}\, V$, $(y,p) \in \mathrm{supp}\,\chi$, the phase function of the expression

$$\exp(-itH_0^h/\hbar)\tilde{\chi}^h(y,D)d_L^h(x)$$

$$= h^{-n}\int \exp(\{i(x-y)\cdot p - itp^2/2\}/\hbar)\chi(y,p)d_L^h(y)dydp$$

satisfies

$$|\partial/\partial p\{(x-y)\cdot p - tp^2/2\}| = |x-y-tp|$$

$$\geq \tfrac{1}{2}(|y|+t|p|) - |x| \geq 2L + \tfrac{t}{4}v_1 - L \geq \tfrac{1}{4}(tv_1+L).$$

Thus integrations by parts in p-variables imply

(4.18) $\quad ||V \exp(-itH_0^h/\hbar)\tilde{\chi}(y,D)d_L^h|| \leq C_N h^{N+1}(1+t)^{-N-1}.$

The relations (4.17) and (4.18) clearly imply the statement of Proposition 4.3.

§5. Stationary phase method, Theorem.

Putting the relations (1.8), (3.4) and (4.15) altogether, we see that for $T_1 = (2a+v_1)/v_1$, $L \geq 3(T_1+a)$, $R \geq 2L$, $T_4 = T_3+T_1$ with T_3 in (4.3)

(5.1)
$$\sup_{t\geq 0} ||e^{-i(t+T_4)H^h/\hbar} e^{iT_1 H^h/\hbar} W_-^h f_R^h - e^{-itH_0^h/\hbar} e^{-iT_4 H^h/\hbar} e^{iT_1 H_0^h/\hbar} f_L^h - e^{-i(t+T_4-T_1)H_0^h/\hbar}(f_R^h - f_L^h)|| \leq C_N h^N.$$

We multiply the functions inside the norm of (5.1) by $\exp(i(t+T_3)H_0^h/\hbar)$, and then take the limit $t \to \infty$. It follows by the definition of S^h that

(5.2)
$$||(S^h - 1)f_R^h - (e^{iT_3 H_0^h/\hbar} e^{-iT_4 H^h/\hbar} e^{iT_1 H_0^h/\hbar} - 1)f_L^h|| \leq C_N h^N.$$

The second term in the norm of (5.2) is independent of R and is a quantity of finite time problem. Hence it is possible to obtain its asymptotic expansion as $h \to 0$ up to any order. (However, for simplicity, we perform the expansion up to the first order only.) We follow the argument in [10] and start with the expansion of $\exp(-iT_1 H_0^h/\hbar)f_L^h$.

Lemma 5.1. There exists a constant $C > 0$ independent of $0 < h \leq 1$ such that

(5.3)
$$||\exp(iT_1 H_0^h/\hbar)f_L^h - \exp(-ix_1^2/2T_1\hbar + i\pi/4)|T_1|^{-1/2} g(-x_1/T_1)w_L(\underline{x})|| \leq Ch.$$

Proof. As in (2.4), $\exp(iT_1 H_0^h/\hbar)f_L^h(x) = \exp(iT_1 K_0^h/\hbar)\mathcal{F}^{h*}g(x_1) \times \exp(iT\underline{K}_0^h/\hbar)w_L(\underline{x})$. Applying the stationary phase ([10], II.p.17) to

$$\exp(iT_1 K_0^h/\hbar)\mathcal{F}^{h*}g(x_1) = h^{-1/2}\int e^{i\{x_1\cdot p_1 + T_1 p_1{}^2/2\}/\hbar} g(p_1)dp_1 ,$$

we see

(5.4)
$$||\exp(iT_1 K_0^h/\hbar)\mathcal{F}^{h*}g(x_1) - \exp(-ix_1^2/2T_1\hbar + i\pi/4)|T_1|^{-1/2} g(-x_1/T_1)|| \leq Ch.$$

On the other hand expanding $e^{iT_1\hbar p^2/2}$ in the Taylor series in the RHS of

$$\exp(iT_1\underline{K}_0^h/\hbar)w_L(\underline{x}) = (2\pi)^{-(n-1)}\int e^{i(\underline{x}-\underline{y})\cdot\underline{p}+iT_1\underline{p}^2\hbar/2} w_L(\underline{y})d\underline{y}d\underline{x} ,$$

we have

$$(5.5) \qquad ||\exp(iT_1K_0^h/\hbar)w_L(\underline{x})-w_L(\underline{x})|| \leq CT_1\hbar||w_L||_{H^2(\mathbb{R}^{n-1})}.$$

The lemma follows from (5.4) and (5.5).

We combine (4.12) with (5.3). It follows

$$(5.6) \qquad ||\exp(-iT_4H^h/\hbar)\exp(iT_1H_0^h/\hbar)f_L^h$$
$$- E_N^h(T)^{\ell}(e^{-ix_1^2/2T_1\hbar+i\pi/4}|T_1|^{-1/2}g(-x_1/T_1)w_L(\underline{x}))|| \leq Ch.$$

By virtue of (4.9) the second member in the LHS of (5.6) may be written as an oscillatory integral

$$(5.7) \; Z^h(T_3)f_L^h(x) = \hbar^{-n\ell/2}\int A_N^h(T_4,x,\theta,y)e^{i\{\phi(T_4,x,\theta,y)-iy_1^2/2T_1\}/\hbar+i\pi/4}$$
$$\times |T_1|^{-1/2}g(-y_1/T_1)w_L(\underline{y})dyd\theta.$$

To investigate (5.7), we need to know the configuration of the Lagrangian manifold $\Omega \equiv \Omega(T_3)D^*$ in the phase space $\mathbb{R}^n \times \mathbb{R}^n$. For a subset $I = \{i_1,\dots,i_\ell\} \subset \{1,2,\dots,n\}$, we denote by π_I the projection from $\mathbb{R}^n\times\mathbb{R}^n$ to $\mathbb{R}^n$ which maps $(x_1,\dots,x_n,p_1,\dots,p_n)$ to $(x_{i_1},\dots,x_{i_\ell},p_{j_1},\dots,p_{j_{n-\ell}})$, $\{j_1,\dots,j_{n-\ell},i_1,\dots,i_\ell\} = \{1,2,\dots,n\}$. By a general theorem about Lagrangian manifolds (Maslov [6]), D may be decomposed into a finite number of open sets: $D \Subset \bigcup_{k=1}^{m} U_k$ such that π_{I_k} is a diffeomorphism on $\Omega(T_3)U_k^*$. We denote by $\mathcal{F}_I^h$ the partial Fourier transform w.r.t. the variables $x_I = (x_{i_1},\dots,x_{i_\ell})$, $I = (i_1,\dots,i_\ell)$

$$(5.8) \qquad (\mathcal{F}_I^hf)(p_I,x_J) = h^{-|I|/2}\int \exp(-ix_I\cdot p_I/\hbar)f(x_I,x_J)dx_I,\ J= I^c.$$

We take $\rho_k(\underline{x},v) \in C_0^\infty(U_k)$ such that $0 \leq \rho_k(\underline{x},v) \leq 1$ and $\sum_{k=1}^{m} \rho_k(\underline{x},v) = 1$ for $(\underline{x},v) \in D$ and set

(5.9) $$f^h_{L,k}(x) = h^{-1/2} \int_{-\infty}^{\infty} \exp(ix_1p_1/\hbar)\rho_k(\underline{x},p_1)g(p_1)w_L(\underline{x})dp_1.$$

We note that on $\Omega(T_3)U_k^*$ one can take (x_{I_k},p_{J_k}) as independent variables and (x_{J_k},p_{I_k}) may be expressed by them; we set, writing $I_k = I$ and $J_k = J$,

(5.10) $$Q^h_k(T_3)f^h_{L,k}(x_I,p_J) \begin{cases} = \exp(-i(\mu_k+\gamma)\pi/2+i(1+|J|)\pi/4+i\lambda_k(x_I,p_J)/\hbar) \\ \quad \times |\det \partial\pi_I\Omega(T_3)/\partial\underline{a}\partial v(\underline{a},v)^*|^{-1/2} \rho_k(\underline{a},v) \\ \quad \times g(v)w_L(\underline{a}), \text{ when } (x_I,p_J) = \pi_I\Omega(T_3)\times \\ \quad (\underline{a},v)^*, \quad ,(\underline{a},v) \in U_k; \\ = 0 \quad \text{otherwise.} \end{cases}$$

where μ_k = dimension of negative subspace for the symmetric matrix $\partial x_J/\partial p_J(x_I,p_J)$, γ = Keller-Maslov index of the trajectory $\{\Gamma(t)\Gamma_0(-T_1)(\underline{a},v)^*: -T_1 \leq t \leq T_3\}$ on $\{\Gamma(t)\Gamma_0(-T)D^*: -T_1 \leq t \leq T_3\}$, $|J|$ = the cardinal number of J and

(5.11) $$\lambda_k(x_I,p_J) = \{\int_{-T_1}^{T_3} \{p(t,\underline{a},v)^2/2-V(x(t,\underline{a},v))\}dt-T_1v^2/2\} - x_J(T_3,\underline{a},v)p_J .$$

Using (5.3) and (5.6), and following the computations of Yajima [10], we obtain

<u>Proposition</u> 5.2. Let $f^h_{L,k}(x)$, $I = I_k$, $J = I_k^c$ and $Q^h_k(T_3)f^h_{L,k}(x_I,p_J)$ be as above. Then there exists a constant $C > 0$ independent of $0 < h \leq 1$ such that

(5.12) $$||\mathcal{F}^h_J Z^h(T_3)f^h_{L,k}(x_I,p_J) - Q^h_k(T_3)f^h_{L,k}(x_I,p_J)|| \leq Ch.$$

In fact we want to have the expansion of $\mathcal{F}^h Z^h(T_3)f^h_L = \sum_{k=1}^m \mathcal{F}^h Z^h(T_3)\times f^h_{L,k} = \sum_{k=1}^m \mathcal{F}^h_{I_k} Q^h_k(T_3)f^h_{L,k} + O(h)$. However we know that the phase function $-x_{I_k}\cdot p_{I_k}+\lambda_k(x_{I_k},p_{J_k})$ in the integral representation of $\mathcal{F}^h_{I_k} Q^h_k(T_3)f^h_{L,k}$ has in general singular Hessians on $\operatorname{supp} Q^h_R(T_3)f^h_{L,k}$ and the method of stationary phase does not apply. In fact it does not in general have the expansion in h. (Although we know the generic forms of such singular $\lambda_k(x_{I_k}, p_{J_k})$ by Arnold's theory [1] and for such generic $\lambda_k(x_{I_k}, p_{J_k})$ it may be possible to give the expansions, we really do not know if our $\lambda_k(x_{I_k}, p_{J_k})$ belongs to these generic classes.)

Thus we restrict our attention only to those p's for which only a finite number of $\{(\underline{a}_j, v_j) \in D, j=1, \ldots, d\}$ satisfy $p = \pi_\phi \Omega(T_3)(\underline{a}_j, v_j)^*$ and $\pi_{\{\phi\}}$ is non-singular at each $\Omega(T_3)(\underline{a}_j, v_j)^*$. We fix one of such p and write it as p_0. It is clear that $p_0 \neq ve_1$ for any $v>0$ and that if $\varepsilon>0$ is small enough $(\pi_\phi \Omega(T_3))^{-1}\{p: |p-p_0|<\varepsilon\} \cap D^*$ is written as $\bigcup_{j=1}^{d} V_j^*$ with disjoint neighbourhoods V_j of (a_j, v_j) and π_ϕ is diffeomorphic on each $\Omega(T_3)V_j^*$. Moreover we may assume that V_j is contained in one of U_k with $I_k = \phi$ which we write U_j and $\rho_j(\underline{x}, v) = 1$ on V_j, changing the choice of the covering $\{U_k\}$ and the partition of unity $\{\rho_k\}$ if necessary. Since $p = (p_1, \ldots, p_n)$ are local coordinates on $\Omega(T_3)V_j^*$, $(x,p) \in \Omega(T_3)V_j^*$ defines a function $x = x(p)$.

<u>Lemma</u> 5.3. If T_3 is large enough, the matrix $\partial x(p)/\partial p$ is positive definite for any $(x,p) \in \Omega(T_3)V_j^*$, $j=1,\ldots,d$.

<u>Proof</u>. Differentiating (4.3), we have

$$\partial x(T_3, \underline{a}, v)/\partial \underline{a}\partial v = (T_3-T_2)\partial p(T_3, \underline{a}, v)/\partial \underline{a}\partial v + \partial x(T_2, a, v)/\partial \underline{a}\partial v \quad .$$

Since $\partial p(T_3, \underline{a}, v)/\partial \underline{a}\partial v$ is non-singular for $(a,v) \in \bigcup_{j=1}^{d} V_j$, we see for $(x(p),p) \in \Omega(T_3)V_j^*$,

$$\partial x(p)/\partial p = (T_3-T_2) + \partial x(T_2, a(p), v(p))/\partial p.$$

this obviously implies the lemma.

We write $p = |p|\hat{p}$, $\hat{p} \in S^{n-1}$. $d\sigma_p$ is the surface element of S^{n-1}. In polar coordinates, $dp = |p|^{n-1} d|p| d\sigma_p$. Thus we have

$$(5.14) \qquad |\det \partial p(T_3, \underline{a}, v)/\partial \underline{a}\partial v| = |p(T_3, a, v)|^{n-1} d|p| d\sigma_p/d\underline{a}\cdot dv$$
$$= v^{n-1} d\sigma_p/d\underline{a}(\underline{a}, v)$$

as $|p(T_3, \underline{a}, v)| = v$.

Applying Lemma 5.3 and (5.14) to (5.10), we see that for $|p-p_0| <$

ε with $v_3+\varepsilon < |p_0| < v_4-\varepsilon$,

$$(5.15)\qquad Q_j^h(T_3)f_{L,j}^h(p) = \exp(-i\pi\gamma_j/2+i(n+1)\pi/4+i\lambda_j(p)/\hbar)$$
$$\times |p|^{-(n-1)/2}(d\sigma_p/d\underline{a})(\underline{a},v)^{-1/2}, \quad p = \pi_\phi\Omega(T_3)(\underline{a},v).$$

Combining (5.6), (5.15) with Proposition 5.2, we obtain the following lemma.

Lemma 5.4. Let p_0 and $\varepsilon>0$ be as above. Then

$$(5.16)\qquad ||w_{\varepsilon/2}(p-p_0)\{\mathcal{F}^h e^{-iT_4H^h/\hbar} e^{iT_1H_0^h/\hbar} f_L^h(p)$$
$$- \sum_{j=1}^{d} e^{-i\pi\gamma_j/2+i(n+1)\pi/4+i\lambda_j(p)/\hbar} |p|^{-(n-1)/2}(d\sigma_p/da)(\underline{a}_j,|p|)^{-1/2}\}||$$
$$\leq Ch,$$

where the summation is taken over all $\underline{a}_j$'s such that $p = \pi_\phi\Omega(T_3)(\underline{a}_j, |p|)^*$.

We multiply the functions inside the norm by $\exp(iT_3p^2/\hbar)$ and then subtract $w_{\varepsilon/2}(p-p_0)\mathcal{F}^h f_L^h$. Since we clearly have

$$(5.17)\qquad ||w_{\varepsilon/2}(p-p_0)\mathcal{F}^h f_L^h|| \leq C_N h^N,$$

(5.2), (5.16) and the unitarity of $\mathcal{F}^h$ imply the following theorem.

Theorem 5.5. Let Assumption (C) be satisfied. Suppose that $p_0 \in \mathbb{R}^n \setminus \{0\}$ be such that (i) there are only finite number of $\underline{a}_j \in \mathbb{R}^{n-1}$ ($j=1,\ldots,d$) such that the solutions $(x(t),p(t))$ of (1.5) with $x(t) = t|p_0|e_1+(0,\underline{a}_j)$ at large negative t satisfy $p(t) = p_0$ at large positive t; (ii) $d\sigma_p/d\underline{a}(\underline{a}_j,|p_0|) \neq 0$. Then there exists a $\varepsilon > 0$ such that

$$||w_\varepsilon(p-p_0)\{h^{(n-1)/2}|p|^{-1}T(p,|p|e_1)$$
$$- \sum_{j=1}^{d} |p|^{-(n-1)/2} e^{-i\pi\gamma_j/2+i(n+1)\pi/4+iS_j(p)/\hbar}$$
$$d\sigma_p/d\underline{a}\ (a_j,|p|)^{-1/2}\}|| \leq Ch$$

where the summation is taken over all a_j such that condition (i) is satisfied, γ_j is the Keller-Maslov-index,

$$S_j(p) = \lim_{\substack{\sigma\to-\infty\\ \tau\to\infty}} \{\int_\sigma^\tau\{p(t,a_j,|p|)^2/2-V(x(t,a_j,|p|))\}dt-p^2\tau/2+p^2\sigma/2\} +$$

$$\lim_{t\to\infty}(x(t,a_j,v)-tp)\cdot p.$$

§6. The classical limit of total cross section.

The total cross section $\sigma^h_{tot}(ke_1)$ for (1.1) with the incoming momentum ke_1 is defined by

$$(6.1)\qquad \sigma^h_{tot}(ke_1) = |k|^{n-3}\int_{S^{n-1}}|T^h(k\omega,ke_1)|^2 d\omega.$$

Recently Enss and Simon [4] showed that $\sigma^h_{tot}(ke_1)$ can also be obtaind as a function $\sigma^h_{tot}(ke_1)$ such that the equation

$$(6.2)\qquad \lim_{R\to\infty}||(S^h-1)\tilde{f}^h_R||^2 = \int_0^\infty \sigma^h_{tot}(ke_1)|g(k)|^2 dk$$

is satisfied for $g \in C_0^\infty((0,\infty))$ and $\tilde{f}^h_R(x) = h^{-(n-1)/2}(\mathcal{F}^{h*}g)(x_1)w_R(\underline{x})$. They also conjectured that the classical limit of the total cross section is twice the classical total cross section, which is by definition the $(n-1)$-dimensional volume of $\underline{a} \in \mathbb{R}^{n-1}$ such that $x(t,\underline{a},k) \neq tke_1+\underline{a}$ for some $t \in \mathbb{R}$:

$$(6.3)\qquad \lim_{h\downarrow 0}\sigma^h_{tot}(ke_1)h^{(n-1)} = 2\sigma^c_{tot}(ke_1).$$

In this section we shall give a sufficient condition for (6.3) to hold.

We take $(v_1,v_2) \subset \mathbb{R}^+$ and split $\mathbb{R}^{n-1}\times(v_1,v_2)$ into

$$\Omega_1 = \{(\underline{a},v)\colon v_1<v<v_2 \text{ and } x(t,\underline{a},v) = tve_1+(0,\underline{a}) \text{ for all } t\},$$

$$\Omega_2 = \{\mathbb{R}^{n-1}\times(v_1,v_2)\}\setminus\Omega_1.$$

<u>Assumption</u> (A) The boundary of Ω_1 is not too wild:

$$L^2(\Omega_1) = \overline{C_0^\infty(\Omega_1)},\quad L^2(\Omega_2) = \overline{C_0^\infty(\Omega_2)}.$$

We take two functions $\sigma(\underline{x},v) \in C_0^\infty(\Omega_2)$ and $\tau(\underline{x},v) \in C_0^\infty(\Omega_1)$. By (4.15) and (6.2), we see with $L \geqq 3(T_1+a)$

$$(6.3)\qquad |h^{(n-1)}\int_0^\infty \sigma^h_{tot}(ke_1)|g(k)|^2 dk - ||(e^{iT_3H_0^h/\hbar}e^{-i(T_3+T_1)H^h/\hbar}$$
$$e^{iT_1H_0^h/\hbar} - 1)f^h_L||^2| \leq C_N h^N .$$

For $\tau \in C_0^\infty(\Omega_1)$ we set

$$f^h_{\tau,L}(x) = h^{-1/2}\int e^{ix_1\cdot p_1/\hbar}\tau(\underline{x},p_1)w_L(\underline{x})g(p_1)dp_1$$

and $f^h_{\sigma,L}(x)$ similarly for $\sigma \in C_0^\infty(\Omega_2)$.

Lemma 6.1. For any $\tau \in C_0^\infty(\Omega_1)$

$$(6.4)\qquad \|(e^{iT_3H_0^h/\hbar}e^{-i(T_3+T_1)H^h/\hbar}e^{iT_1H_0^h/\hbar} - 1)f^h_{\tau,L}\| \le C_N h$$

Proof. It is clear that $\pi_{\{1,..,n\}}$ is diffeomorphic on $\Omega(T_3)(\mathrm{supp}\,\tau)^*$ and by Proposition 5.2, $\exp(-i(T_3+T_1)H^h/\hbar)\exp(iT_1H_0^h/\hbar)f^h_{\tau,L}$ is approximated by the RHS of (5.10) with $I_k = \{1,2,...,n\}$, $\mu_k=0$, $\gamma_k=1$, $|J|=0$, $a=\underline{x}$, $v=x_1/T_3$, $|\det \partial\pi_I\Omega(T_3)/\partial\underline{a}\partial v(\underline{a},v)^*| = T_3$, $\lambda(x) = \{\int_{-T_1}^{T_3}\{(v^2/2\}dt - T_1v^2/2\} = T_3v^2/2 = x_1^2/2T_3$ and $\rho_k = \tau$, that is,

$$(6.5)\qquad \|\exp(-i(T_3+T_1)H^h/\hbar)\exp(iT_1H_0^h/\hbar)f^h_{\tau,L}(x) - \sqrt{T_3}^{-1}\exp(ix_1^2/2T_3\hbar - i\pi/4)\tau(\underline{x},x_1/T_3)g(x_1/T_3)w_L(\underline{x})\| \le Ch.$$

On the other hand the computation as in Lemma 5.1 shows

$$(6.6)\qquad \|\exp(-iT_3H_0^h/\hbar)f^h_{\tau,L}(x) - \sqrt{T_3}^{-1}\exp(ix_1^2/2T_3\hbar - i\pi/4)\tau(\underline{x},x_1/T_3)g(x_1/T_3)w_L(\underline{x})\| \le Ch.$$

(6.5), (6.6) and the unitarity of $\exp(iT_3H_0^h/\hbar)$ clearly imply (6.4).

Since $\Omega_2 \subset D$, we have $\sigma(\underline{x},v) = \Sigma\sigma(\underline{x},v)\rho_k(\underline{x},v)$, using the partition of unity in Sect.5, and consequently

$$(6.7)\qquad f^h_{\sigma,L}(x) = \sum_{k=1}^{m} f^h_{\sigma_k,L}(x)\ ,\quad \sigma_k(\underline{x},v) = \sigma(\underline{x},v)\rho_k(\underline{x},v).$$

Lemma 6.2. If $J_k \cap \{2,3,...,n\} \neq \phi$, $J_k = I_k^c$, then

$$(6.8)\qquad \lim_{h\to 0}(e^{-i(T_3+T_1)H^h/\hbar}e^{iT_1H_0^h/\hbar}f^h_{\sigma_k,L},\ e^{-iT_3H_0^h/\hbar}f_{\sigma,L}) = 0$$

Proof. By Proposition 5.2, and Plancherel theorem, we see that the LHS of (6.8) is equal to

$$(6.9)\qquad \lim_{h\downarrow 0}(Q_k^h(T_3)f^h_{\sigma_k,L},\ \mathcal{F}^h_{J_k}e^{-iT_3H_0^h/h}f^h_\sigma).$$

When $J_k \cap \{2,3,...,n\} \neq \phi$, say $2 \in J_k$, it is easy to see that for any $\varepsilon > 0$,

$$(6.10)\qquad \int dx_{I_k}\int dp_{J_k\setminus\{2\}}\int_{|p_2|>\varepsilon}|\mathcal{F}^h_{J_k}e^{-iT_3H_0^h/\hbar}f^h_{\sigma_k,L}(x_{I_k},p_{J_k})|^2dp_2$$

$$= \int_{|p_2|>\varepsilon} dp_2 \int dp_1 dx_3 \ldots dx_n \Big| h^{-1/2} \int_{-\infty}^{\infty} e^{-ix_2 \cdot p_2/\hbar} \sigma_k(\underline{x}, p_1) g(p_1) w_L(\underline{x}) dx_2 \Big|^2$$

$$\leq C_N h^N , \quad 0 < h \leq 1$$

for any $N > 0$. From (6.10), it follows that (6.9) vanishes.

To treat the case when $J_k = \phi$ and $J_k = \{1\}$, we need Riemann-Lebesgue theorem of the following form.

Lemma 6.3. Let $\alpha(x)$ be a real valued C^1-function and $f(x) \in L^1(\mathbb{R}^n)$. Suppose that grad $\alpha(x) \neq 0$ almost everywhere. Then

$$\lim_{h\to 0} \int e^{i\alpha(x)/\hbar} f(x) dx = 0 .$$

By virtue of Lemma 6.3, when $I = \{1,2,\ldots,n\}$

$$\lim_{h\to 0} (Q_k^h(T_3) f^h_{\sigma_k,L}, e^{-iT_3 H_0^h/\hbar} f^h_\sigma) = 0 ,$$

unless there is an open set $G \subset \{\mathrm{supp}\ \sigma(\underline{x}, x_1/T_3) w(\underline{x}) g(x_1/T_3)\} \cap \mathrm{supp}\ Q_k^h(T_3) f^h_{\sigma_k,L}$ on which $\nabla\lambda_k(x) = (x_1/T_2, 0)$, i.e. unless there are two open set F_1 and $F_2 \subset \Omega_2$ such that for any $(a_1, v_1) \in F_1$ there exists $(a_2, v_2) \in F_2$ such that

$$(6.11) \qquad x(T_3, a_1, v_1) = (T_3 v_2, a_2), \ p(T_3, a_1, v_1) = v_2.$$

Likewise, when $J_k = \{1\}$,

$$\lim_{h\to 0} (Q_k^h(T_3) f^h_{\sigma_k,L}, \mathcal{F}^h_{\{1\}} e^{-iT_3 H_0^h/\hbar} f^h_\sigma)$$

$$= \lim_{h\to 0} (Q_k^h(T_3) f^h_{\sigma_k,L}, e^{-iT_3 p_1^2/\hbar} \sigma(\underline{x}, p_1) g(p_1) w_L(\underline{x}))$$

and this quantity vanishes unless the same situation as in the case $J = \phi$ occurs.

Assumption (B) (No interference condition of incoming and outgoing waves). There are no pairs of open sets F_1 and $F_2 \subset \Omega_2$ such that

$$(6.12) \qquad \Gamma(T_3+T_1)\Gamma_0(-T_1) F_1^* = \Gamma_0(T_3) F_2^* .$$

Theorem 6.4. Suppose that Assumptions (A), (B) and (C) are satisfied. Then for any $g \in C_0^\infty(\mathbb{R}^+)$

$$(6.13)\qquad \lim_{h\downarrow 0} h^{(n-1)} \int_0^\infty \sigma^h_{tot}(ke_1)|g(k)|^2 dk$$

$$= 2\int_0^\infty \{\int_{\mathbb{R}^{n-1}} \chi_{\Omega_2}(\underline{x},k)d\underline{x}\}\ |g(k)|^2 dk,$$

where χ_{Ω_2} is the characteristic function of Ω_2.

Proof. Let $\tau \in C_0^\infty(\Omega_1)$ and $\sigma \in C_0^\infty(\Omega_2)$. By (6.4), the unitarity of propagators and Assumption (A), we have

$$(6.14)\qquad \lim_{h\to 0} ||(e^{iT_3H_0^h/\hbar} e^{-i(T_3+T_1)H^h/\hbar} e^{iT_1H_0^h/\hbar} - 1)f^h_{\chi_{\Omega_1},L}||^2 = 0 .$$

On the other hand, Proposition 5.2. Lemma 6.2 and argument above imply that under Assumption (B),

$$\lim_{h\to 0} (e^{-i(T_3+T_1)H^h/\hbar} e^{iT_1H_0^h/\hbar} f^h_{\sigma,L}, e^{-iT_3H_0^h/\hbar} f^h_{\sigma,L}) = 0$$

, that is, that

$$(6.15)\qquad \lim_{h\to 0}||(e^{iT_3H_0^h/\hbar} e^{-i(T_3+T_1)H^h/\hbar} e^{iT_1H_0^h/\hbar} -1)f^h_{\sigma,L}||^2 = 2||f^h_{\sigma,L}||^2.$$

Then by Assumption (A) and the unitarity of propagators,

$$(6.16)\qquad \lim_{h\downarrow 0} ||(e^{iT_3H_0^h/\hbar} e^{-i(T_3+T_1)H^h/\hbar} e^{iT_1H_0^h/\hbar} -1)f^h_{\chi_{\Omega_2}}||^2 = 2||f^h_{\chi_{\Omega_2}}||^2$$

$$= 2\int_0^\infty dk \int_{\mathbb{R}^{n-1}} |\chi_{\Omega_2}(\underline{x},k)||g(k)|^2 d\underline{x} .$$

(6.3), (6.14) and (6.15) obviously imply the conclusion of the theorem.

References.

1. Arnold, V.I., Mathematical methods of classical mechanics, Springer, New-York-Berlin-Heidelberg, 1978.
2. Asada, K. and D. Fujiwara, On some oscillatory integral transformations in $L^2(R^n)$, Japan J. Math. 4 (1978), 299-361.
3. Fujiwara, D., A construction of the fundamental solution for the Schrodinger equation, J. d'Analyse Math. 35 (1979), 41-96.
4. Enss, V. and B. Simon, Finite total cross sections in nonrelativistic quantum mechanics, Commun. Math. Phys. 76 (1980), 177-209.
5. Kuroda, S.T., An introduction to scattering theory, Lecure notes series 51 (1978), Mat. Institut, Aarhus University, Aarhus.
6. Maslov, V.P., Theorie des perturbations et methodes asymptotiques, Dunod, Paris 1972.
7. Miller, W.H., Classical-limit quantum mechanics and the theory of molecular collisions, Adv. Chem. Phys. 25 (1974), 69-177.
8. Simon, B., Quantum mechanics for Hamiltonians defined as quadratic forms, Princeton Univ. Press, Princeton N.J. 1971.
9. Vainberg, B.R., Quasi-classical approximation in stationary scattering problems, Funct. Anal. Appl. 11 (1977), 6-18 (Russian).
10. Yajima, K., The quasi-classical limit of quantum scattering theory, I and II, Commun. math. Phys. 69 (1979), 101-129 and Duke J. Math. 48 (1981), 1-22.
11. Yajima, K., The quasi-classical approximation to Dirac theory I and II, J. Fac. Sci. Univ. Tokyo 29 (1982), 161-194 and 371-386.
12. Yajima, K., The quasi-classical limit of scattering amplitude II, general potentials, in preparation.

FONDAZIONE C.I.M.E.
CENTRO INTERNAZIONALE MATEMATICO ESTIVO
INTERNATIONAL MATHEMATICAL SUMMER CENTER

"Probability and Analysis"

is the subject of the First 1985 C.I.M.E. Session.

The Session, sponsored by the Consiglio Nazionale delle Ricerche and the Ministero della Pubblica Istruzione, will take place under the scientific direction of Prof. GIORGIO LETTA and Prof. MAURIZIO PRATELLI (Università di Pisa, Italy) at «Villa Monastero», Varenna (Como), Italy, ***from May 31 to June 8, 1985.***

Courses

a) ***Some Applications of Probability to Geometry*** (6 lectures in English).
Prof. Jean-Michel BISMUT (Université Paris-Sud).

— Large deviations and the asymptotics of the heat kernel.
— The index theorem of Atiyah-Singer for Dirac operators.
— Some geometrical properties of the loop space of a Riemannian manifold.
— The Witten complex and its applications.

References

- M.F. Atiyah: Classical groups and classical operators on manifolds. In «Differential operators on manifolds». CIME, Cremonese, Roma 1975.
- M.F. Atiyah, I.M. Singer: The index of elliptic operators III. Ann. of Math. 87 (1968), 546-604.
- J.M. Bismut: Large deviations and the Malliavin calculus. Progress in Math. n. 45, Birkhauser, 1984.
- B. Doubrovine, S. Novikov, A. Fomenko: Géométrie contemporaine I and II. Editions MIR (also English edition by Springer) (see the basics on connections in these two books).
- WITTEN E.: Supersymmetry and Morse theory. J. of Diff. Geom. 17 (1982), 661-692.

b) ***Martingales and Fourier Analysis in Banach Spaces.*** (6 lectures in English).
Prof. Donald L. BURKHOLDER (University of Illinois, Urbana).

— Martingale transforms and the geometry of Banach spaces.
— A geometrical condition implying the boundedness of singular integral operators.
— Bourgain's converse.
— Optimal control of martingales.
— Boundary value problems and sharp inequalities for martingale transforms and stochastic integrals.
— Some other problems and recent progress in probability, geometry, and Fourier Analysis.

References

- J. BOURGAIN, Some remarks on Banach spaces in which martingale difference sequences are unconditional. Ark. Mat. 21 (1983), 163-168.
- D.L. BURKHOLDER, A Geometrical characterization of Banach spaces in which martingale difference sequences are unconditional. Ann. Probab. 9 (1981), 997-1011.
- D.L. BURKHOLDER, A geometric condition that implies the existence of certain singular integrals of Banach-space-valued functions, in «Conference on Harmonic Analysis in Honor of Antoni Zygmund» (W. Beckner, A.P. Calderon, R.F. Fefferman, and P.W. Jones, editors), Wadsworth, Belmont, California, 1983.
- D.L. BURKHOLDER, Boundary value problems and sharp inequalities for martingale transforms. Ann. Probab. 12 (1984), 647-702.

c) ***Martingale Theory: an Analytical Formulation with some Applications in Analysis***. (6 lecture in English).
Prof. S. D. CHATTERJI (EPF Lausanne).

— A basic convergence theorem: its statement and discussion. Its relation to martingale, semimartingale et amart convergence theory.
— Proof of the convergence theorem.
— Applications:

(a) Lebesgue differentiation theorem
(b) Product measures
(c) Ryll-Nardzewski's fixed point theorem
(d) Miscellaneous.

References

- S.D. CHATTERJI, Les martingales et leurs applications analytiques. Lecture Notes in Mathematics, vol. 307, Springer Verlag, Berlin, 1973.

d) ***Probabilistic Methods in the Geometry of Banach Spaces***. (6 lectures in English).
Prof. Gilles PISIER (Université Paris VI).

Summary

The course will present the notions of type and cotype of a Banach space together with their geometric interpretations. The duality problem between type and cotype will be discussed and the main results related to the notion of k-convexity will be presented. Moreover, the relations of these notions with the famous theorem of Dvoretzky on the spherical sections of convex bodies will be discussed, using some recent inequalities concerning vector valued Gaussian random variables. The latter inequalities can be advantageously used instead of the isoperimetric inequality on the sphere.

References

- B. MAUREY, G. PISIER, Séries de variables aléatoires vectorielles indépendantes et propriétés géométriques des espaces de Banach. Studia Math. 58 (1976), 45-90.
- G. PISIER, Holomorphic semi-groups and the geometry of Banach spaces. Annals of Math. 115 (1982), 375-392.
- T. FIGIEL, J. LINDENSTRAUSS, V. MILMAN, The dimension of the spherical sections of convex bodies. Acta Math. 139 (1977), 53-94.
- Y. BENYAMINI, Y. GORDON, Random factorization of operators between Banach spaces. Journal Analyse. Jerusalem. 39 (1981), 45-74.

Seminars

A number of seminars and special lectures will be offered during the Session.

FONDAZIONE C.I.M.E.
CENTRO INTERNAZIONALE MATEMATICO ESTIVO
INTERNATIONAL MATHEMATICAL SUMMER CENTER

"Some Problems in Nonlinear Diffusion"

is the subject of the Second 1985 C.I.M.E. Session.

The Session, sponsored by the Consiglio Nazionale delle Ricerche and the Ministero della Pubblica Istruzione, will take place under the scientific direction of Prof. ANTONIO FASANO and Prof. MARIO PRIMICERIO (Università di Firenze, Italy) at Villa «La Querceta», Montecatini Terme (Pistoia), Italy, ***from June 10 to June 18, 1985.***

Courses

a) ***The Porous Medium Equation.*** (8 lectures in English).
Prof. D. G. ARONSON (University of Minnesota, USA).

— Physical background, elementary properties, basic existence and uniqueness theory.
— Regularity in one space dimension.
— Properties of the interface in one space dimension: smoothness, nonsmoothness, waiting time.
— The relationship between the porous medium pressure equation and a Hamilton-Jacobi equation: limiting behavior as $m \to 1$.
— Regularity in R^d for $d > 1$: Hölder continuity and quasi convexity.
— Initial trace, existence and uniqueness in R^d: Harnack-type estimates and growth at infinity.
— Asymptotic behavior of solutions of the initial-boundary value problem.
— Stabilization results.

References

- D.G. ARONSON and P. BENILAN, Regularité des solutions de l'équation des milieux poreux dans R^n. C.R. Acad. Sci. Paris, Série A-B, 288 (1979), 103-105.
- D.G. ARONSON and L.A. CAFFARELLI, The initial trace of a solution of the porous medium equation. Trans. Amer, Math. Soc. 280 (1983), 351-366.
- D.G. ARONSON, L.A. CAFFARELLI and J.L. VAZQUEZ, Interfaces with a corner point in one-dimensional porous medium flow. Comm. Pure Appl. Math., to appear.
- D.G. ARONSON, M.G. CRANDALL and L.A. PELETIER, Stabilization of solutions of a degenerate nonlinear diffusion problem. Nonlinear Analysis 6 (1982), 1002-1022.
- D.G. ARONSON, S. KAMIN and L.A. CAFFARELLI, How an initially stationary interface begins to move in porous medium flow. SIAM J. Math. Anal. 14 (1983), 639-658.
- D.G. ARONSON and L.A. PELETIER, Large time behaviour of solutions of the porous medium equation. J. Diff. Eq. 39 (1981), 378-412.
- P. BENILAN, M.G. CRANDALL and M. PIERRE, Solutions of the porous medium equation in R^n under optimal conditions on initial values. Indiana Univ. Math. J. 33 (1984), 51-87.
- L.A. CAFFARELLI and A. FRIEDMAN, Continuity of the density of a gas flow in a porous medium. Trans. Amer. Math. Soc. 252 (1979), 99-113.
- L.A. CAFFARELLI and A. FRIEDMAN, Regularity of the free boundary of a gas flow in an n-dimensional porous medium. Indiana U. Math. J. 29 (1980), 361-391.
- B.E.J. DAHLBERG and C.E. KENIG, Nonnegative solutions of the porous medium equation. Comm. P.D.E., to appear.
- B.F. KNERR, The porous medium equation in one dimension. Trans. Amer. Math. Soc. 234 (1977), 381-415.
- O.A. OLEINIK, S.A. KALASHNIKOV and CHZOU YUI-LIN, The Cauchy problem and boundary problems for equations of the type of nonstationary filtration. Izv. Akad. Nauk SSSR Ser. Mat., 22 (1958), 667-704.

b) ***Qualitative Methods in Reaction-Diffusion Equations.*** (8 lectures in English).
Prof. Joel SMOLLER (University of Michigan).

Outline.

Topological Methods, the Conley Index, Applications to Reaction-Diffusion Equations, the Fitz-Hugh Nagumo equations, Stability and Bifurcation of Stationary Solutions to Predator-Prey Equations, Symmetry-Breaking Solutions of Semilinear Elliptic equations.

References

• J. SMOLLER, Shock Waves and Reaction-Diffusion Equations. Springer Verlag, 1983.

c) ***Reaction-Diffusion Problems in Chemistry.*** (8 lectures in English).
Prof. I. STAKGOLD (University of Delaware).

— The quasilinear parabolic system of equations for the concentration and temperature of a simple, irreversible reaction.
— The Frank-Kamenetskii, Semenov and Gelfand approximations, The notion of criticality.
— The method of upper and lower solutions.
— Existence, uniqueness and multiplicity of solutions.
— The case of nonlinearities that are not piecewise smooth. Extinction, dead cores and quenching.
— Free boundary problems and related estimates.
— Two-phase problems, particularly gas-solid interactions.
— The pseudo-steady-state approximation and estimates for the conversion.

Seminars

A number of seminars and special lectures will be offered during the Session.

FONDAZIONE C.I.M.E.
CENTRO INTERNAZIONALE MATEMATICO ESTIVO
INTERNATIONAL MATHEMATICAL SUMMER CENTER

"Theory of Moduli"

is the subject of the Third 1985 C.I.M.E. Session.

The Session, sponsored by the Consiglio Nazionale delle Ricerche and the Ministero della Pubblica Istruzione, will take place under the scientific direction of Prof. EDOARDO SERNESI (Università di Roma, Italy) at Villa «La Querceta», Montecatini Terme (Pistoia), Italy, ***from June 21 to June 29, 1985.***

Courses

a) ***Moduli of algebraic surfaces.*** (8 lectures in English).
Prof. Fabrizio CATANESE (Università di Pisa, Italy).

— Deformations of comples structures (theory of Kodaira-Spencer-Kuranishi).
— Topological and complex analytic invariants of surfaces.
— Outline of the Enriques-Kodaira classification of complex surfaces.
— Existence of moduli spaces for algebraic varieties.
— Surfaces of general type and their moduli spaces.
— Moduli via periods.

References

- W. BARTH, C. PETERS, A. van de VEN, Compact complex surfaces. Springer Ergebnisse 4, (1984), Berlin-Heidelberg.
- A. BEAUVILLE, Surfaces algébriques complexes. Astérisque 54 (1978).
- E. BOMBIERI, Canonical models of surfaces of general type. Publ. Scient. I.H. E.S. 42 (1973), 447-495.
- F. CATANESE, On the moduli spaces of surfaces of general type. J. Diff. Geom. 19 (1984), 483-515.
- F. CATANESE, Commutative algebra methods and equations of regular surfaces, in «Algebraic Geometry», Proc. Bucharest 1982, Springer L.N.M. 1056 (1983), 68-111.
- D. GIESEKER, Global moduli for surfaces of general type. Inv. Math. 43 (1977), 233-282.
- K. KODAIRA, Collected papers, vol. II.
- K. KODAIRA, J. MORROW, Complex Manifolds, Holt Rinehart Winston (1971), New York.
- D. MUMFORD, Geometric invariant theory. (2nd edition 1982, with J. FOGARTHY as coauthor). Heidelberg, Springer, 1965.
- H. POPP, Moduli theory and classification theory of algebraic varieties. Springer L.N.M. 620 (1977).
- I. SHAFAREVITCH et al., Algebraic surfaces, Proc. Steklov Inst., A.M.S. Translation (1967), Providence.

b) ***The Torelli and Schottky Problems***. (8 lectures in English).
Prof. Ron DONAGI (Northeastern University, Boston).

Outline.

The theory of Prym varieties. Survey of works of Andreotti-Mayer, Beauville, R. Smith, Schottky-Young, van Geemen, Arbarello-De Concini and Shiota. Recent contributions to infinitesimal variations of Hodge structures.

c) ***Cohomology of the Moduli Space of Curves***. (8 lecture in English).
Prof. John HARER (University of Maryland).

— Cohomological dimension of the mapping class group.
— Euler characteristic of the moduli space.
— Stability theorem.
— Survey of work of J. Harer, E. Miller, R. Lee, R. Charney, S. Wolpert.

LIST OF C.I.M.E. SEMINARS

		Publisher
1954 -	1. Analisi funzionale	C.I.M.E.
	2. Quadratura delle superficie e questioni connesse	"
	3. Equazioni differenziali non lineari	"
1955 -	4. Teorema di Riemann-Roch e questioni connesse	"
	5. Teoria dei numeri	"
	6. Topologia	"
	7. Teorie non linearizzate in elasticità, idrodinamica, aerodinamica	"
	8. Geometria proiettivo-differenziale	"
1956 -	9. Equazioni alle derivate parziali a caratteristiche reali	"
	10. Propagazione delle onde elettromagnetiche	"
	11. Teoria della funzioni di più variabili complesse e delle funzioni automorfe	"
1957 -	12. Geometria aritmetica e algebrica (2 vol.)	"
	13. Integrali singolari e questioni connesse	"
	14. Teoria della turbolenza (2 vol.)	"
1958 -	15. Vedute e problemi attuali in relatività generale	"
	16. Problemi di geometria differenziale in grande	"
	17. Il principio di minimo e le sue applicazioni alle equazioni funzionali	"
1959 -	18. Induzione e statistica	"
	19. Teoria algebrica dei meccanismi automatici (2 vol.)	"
	20. Gruppi, anelli di Lie e teoria della coomologia	"
1960 -	21. Sistemi dinamici e teoremi ergodici	"
	22. Forme differenziali e loro integrali	"
1961 -	23. Geometria del calcolo delle variazioni (2 vol.)	"
	24. Teoria delle distribuzioni	"
	25. Onde superficiali	"
1962 -	26. Topologia differenziale	"
	27. Autovalori e autosoluzioni	"
	28. Magnetofluidodinamica	"
1963 -	29. Equazioni differenziali astratte	"
	30. Funzioni e varietà complesse	"
	31. Proprietà di media e teoremi di confronto in Fisica Matematica	"

1964 -	32. Relatività generale	C.I.M.E.
	33. Dinamica dei gas rarefatti	"
	34. Alcune questioni di analisi numerica	"
	35. Equazioni differenziali non lineari	"
1965 -	36. Non-linear continuum theories	"
	37. Some aspects of ring theory	"
	38. Mathematical optimization in economics	"
1966 -	39. Calculus of variations	Ed. Cremonese, Firenze
	40. Economia matematica	"
	41. Classi caratteristiche e questioni connesse	"
	42. Some aspects of diffusion theory	"
1967 -	43. Modern questions of celestial mechanics	"
	44. Numerical analysis of partial differential equations	"
	45. Geometry of homogeneous bounded domains	"
1968 -	46. Controllability and observability	"
	47. Pseudo-differential operators	"
	48. Aspects of mathematical logic	"
1969 -	49. Potential theory	"
	50. Non-linear continuum theories in mechanics and physics and their applications	"
	51. Questions of algebraic varieties	"
1970 -	52. Relativistic fluid dynamics	"
	53. Theory of group representations and Fourier analysis	"
	54. Functional equations and inequalities	"
	55. Problems in non-linear analysis	"
1971 -	56. Stereodynamics	"
	57. Constructive aspects of functional analysis (2 vol.)	"
	58. Categories and commutative algebra	"
1972 -	59. Non-linear mechanics	"
	60. Finite geometric structures and their applications	"
	61. Geometric measure theory and minimal surfaces	"
1973 -	62. Complex analysis	"
	63. New variational techniques in mathematical physics	"
	64. Spectral analysis	"

Year	No.	Title		Publisher
1974 -	65.	Stability problems		Ed. Cremonese, Firenze
	66.	Singularities of analytic spaces		"
	67.	Eigenvalues of non linear problems		"
1975 -	68.	Theoretical computer sciences		"
	69.	Model theory and applications		"
	70.	Differential operators and manifolds		"
1976 -	71.	Statistical Mechanics		Ed. Liguori, Napoli
	72.	Hyperbolicity		"
	73.	Differential topology		"
1977 -	74.	Materials with memory		"
	75.	Pseudodifferential operators with applications		"
	76.	Algebraic surfaces		"
1978 -	77.	Stochastic differential equations		"
	78.	Dynamical systems		Ed. Liguori, Napoli and Birkhäuser Verlag
1979 -	79.	Recursion theory and computational complexity		Ed. Liguori, Napoli
	80.	Mathematics of biology		"
1980 -	81.	Wave propagation		"
	82.	Harmonic analysis and group representations		"
	83.	Matroid theory and its applications		"
1981 -	84.	Kinetic Theories and the Boltzmann Equation	(LNM 1048)	Springer-Verlag
	85.	Algebraic Threefolds	(LNM 947)	"
	86.	Nonlinear Filtering and Stochastic Control	(LNM 972)	"
1982 -	87.	Invariant Theory	(LNM 996)	"
	88.	Thermodynamics and Constitutive Equations	(LN Physics 228)	"
	89.	Fluid Dynamics	(LNM 1047)	"
1983 -	90.	Complete Intersections	(LNM 1092)	"
	91.	Bifurcation Theory and Applications	(LNM 1057)	"
	92.	Numerical Methods in Fluid Dynamics	(LNM 1127)	"
1984	93.	Harmonic Mappings and Minimal Immersions	to appear	"
	94.	Schrödinger Operators	(LNM 1159)	"
	95.	Buildings and the Geometry of Diagrams	to appear	"
1985 -	96.	Probability and Analysis	to appear	"
	97.	Some Problems in Nonlinear Diffusion	to appear	"
	98.	Theory of Moduli	to appear	"

Note: Volumes 1 to 38 are out of print. A few copies of volumes 23,28,31,32,33,34,36,38 are available on request from C.I.M.E.